DNA Anomalies

By Steve Preston

1st Edition

© Copyright 2017, Steve Preston
All rights reserved. Updated 2018
No part of this book may be reproduced, stored in a retrieval system, or transmitted by any means, electronic, mechanical, photocopying, recording, or otherwise, without written permission from the author.

Table of Contents

DNA Anomalies 1
Table of Contents 3
Introduction 5
DNA Mutation Tracking 13
Mitochondria Anomaly 26
Haplotyping 31
Haplogrouping 33
Y-Chromosome Haplotypes 36
Mitochondrial Haplotypes 42
Staying in Africa Anomaly 45
Monogenism Anomaly 47
Short Overview of Modern Timing 50
More DNA Anomaly 69
Race Mutation Anomaly 75
Stone Bone Anomaly Cretaceous Age 77
End Of The Dinosaurs Anomaly 80
Homo-Gigantus and Capensis Race Anomaly 95
Unclean Anomaly 101
Evolution Anomalies 106
Squid Anomalies 115
Neanderthal DNA Anomaly 120
Holocene Giants Anomaly 10,000 Years Ago 127
Giant Lineage Anomaly 138
Two Types of Humans Anomaly 140
Denisovan Anomaly 142
Homo Capensis Inbreeding Anomaly 150
Huge Brain Anomaly 151
X- Mutation Anomaly 155
Haplogrouping by Y-DNA 162
Haplogrouping by mtDNA 165
Pleistocene Extinction and Survivors [7000BC] 167
Bharata War Anomaly 170
Chimp and Human Differences 178
Vanara Mutation 193
Where Are the Ape-men Now? 204

Homo Sapiens Cognatus 208
Ubaid People Anomaly 213
Considerations 228
About the Author 231

Introduction

I know what you are thinking! How can there be anomalies in DNA? It is, essentially, ones and zeros made from 4 different nucleotides holding a bunch of sugar together. As shown below, these are Adenine, Thymine, Cytosine, and Guanine. To make it even less confusing, Adenine always attaches to Thymine and Cytosine always attaches to Guanine. It can't be faked or misunderstood unless someone really doesn't like the answers they get; as I have brought out in the other anomaly books. Once that happens, the gloves are off and anything is ok. Those calling themselves scientists twist every bit of truth out of what they know to present a "consensus" theory that a group of "special" people can use to mold the minds of the populace.

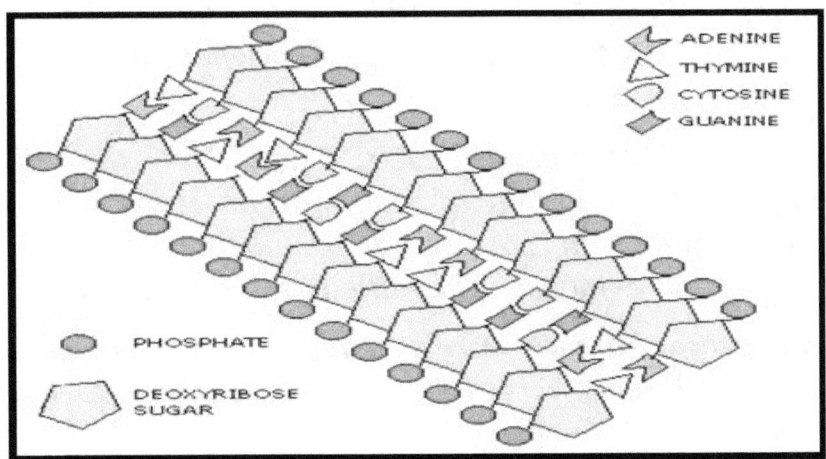

Here is the problem. There are many things listed as anomalies and are forgotten so that one theory or another can be preserved or some group of people is comforted in their ignorance. If you do not want to be comforted and wish to learn about evidence that has taken on the "ANOMALY" classification, please review this book. I can't ensure that the details in this book will make you comfortable, but your eyes will be opened on the following topics.

DNA tells us Humans Did Not Come Out of Africa-*DNA scientists now know people did not come out of Africa as some are still insisting. This presents a problem in evolution so the data is hidden away.*

Only Two Major DNA Mutation Events-*DNA Haplotyping has determined there were 2 massive mutations of human's DNA. The first was about 10 thousand years ago and a second one was about 5500 years ago. Quasi-scientists resist the idea that something happened during both of these critical times as if our DNA just broke down for no reason.*

Humans Went Back to Africa *after the Pleistocene Extinction- DNA Tracking shows that when some people returned; they stayed.*

Homo-Erectus and Ergaster Were Not the First Humans-*The evolution of man was not even close to what you were told. To keep you from being confused, all the 'unusual' evidence is identified as anomaly.*

Neanderthal humans were smarter *than modern humans and their brains were 10% larger according to many studies.*

Neanderthal and Paracas humans had Alien DNA. That being said, they are not related to green men from outer space as many have presented.

Historical Evidence of Lizard Skinned People living between 3100 and 2000BC should be presented when establishing a DNA tracing theory based on known events. If not, we will certainly foster a lie by omission.

Americans Did Not Come from Asia-The land-bridge "theory" presented in schools is a made-up fantasy easily proven wrong by DNA and common sense.

Homo Gigantus People Lived During the Mesozoic- These highly civilized humans wore shoes, made batteries, established and used 16 nuclear processing plants, and modified DNA. This detail messes up the "theory" of evolution so many hundreds of pieces of physical evidence are, essentially, erased from our history. Yes; ancient people walked with dinosaurs and there is proof.

Theban Egyptians were not African as proven by DNA. Instead, they were Semitic.

Neanderthal were red-headed and had tenor voices according to DNA. They also had some type of language which did not include different levels of grunting as some have presented.

Denisovan human DNA is found mostly in Melanesia but, Denisovans have similar DNA to Neanderthal who came from Europe and a Denisovan bone was also found in northern Russia. The DNA map for this race is almost a joke, but may give us needed clues.

Neanderthal had no African DNA mutations *so he could not have originally come from Africa.*

Our brain got smaller *5500 years ago so Cro-Magnon brains were much larger than our modern brains. While this is known, it is classified as anomaly rather than having to discuss what must have happened.*

Ape-People*-Another thing you should know is that Ape-like people that lived a few thousand years ago were, evidently, very common. They are sometimes called Venera people. You probably didn't even know about them and think it is absurd, but the evidence tells a different story. Comfort historians hate the idea of this type of horrible mutation during a war that they choose to ignore. Therefore, all the data is identified as anomaly.*

Bonobo Mutation*-We now know by DNA analysis that Chimpanzees and Bonobos descended from man about 6 thousand years ago, but evolutionist fear that their premise would be in jeopardy if people recognized the strangeness. Therefore, they try to make the data disappear. Unfortunately for them, there is too much.*

Pleistocene War*- According to nuclear residue, mutation levels, melted walls, and many other things impossible to ignore, there was a Nuclear War during the Pleistocene that caused major mutations of humans. With hundreds of pieces of evidence, one would think this would be factored into our history, but all of it is almost always ignored.*

Bharata War*-There was a second horrible nuclear war before World War II that was more destructive than any* war since. While there is an enormous amount of evidence about this war, the thing important here, is that it was responsible for massive mutations of humans and most historians and anthropologists are lying to you.

Snakes did not lose their legs in the Garden of Eden, but still it is described as dogma in religious ceremonies even though the Bible doesn't say snakes had legs. The man who seduced Eve in the Garden was well described by Moses and many other ancient historians, but people didn't like his capabilities so they turned him into a snake.

Unclean Animals-Ancient Geneticists developed all sorts of oddball animals called "<u>unclean</u> abominations" by Moses in the book of Genesis were abominations because they had their DNA modified, not because God hated animals he had created. This is just stupid, but it does keep people from having to describe the manipulation of DNA that was common and described in many, many ancient texts.

Remade Dinosaurs- We know Pleistocene geneticists redeveloped dinosaurs about 20 thousand years ago. This is because we are finding hundreds of dinosaur remains that are not fossilized. Because "quasi-scientists' don't like to accept that civilized humans, described in many ancient texts, used advanced scientific methods and experimentation.

Speaking of having a larger brain- Some try to say we are smarter today than Neanderthal, Homo-Capensis, and Cro-Magnon who all had larger brains. Additionally, they ignore all the evidence that 5500 years ago our brains began to atrophy; we lost the ability of understanding multiple languages; and lost perceptions recorded around the world. All we need to do is look at how our brain has shrunk.

Substantial Neanderthal DNA is located in South America- Wait a minute. Weren't we told Neanderthal were all in Europe?

Base Composition of DNA				
Organism	Percentage of base in organism's DNA			
	adenine (%)	guanine (%)	cytosine (%)	thymine (%)
Maize	26.8	22.8	23.2	27.2
Octopus	33.2	17.6	17.6	31.6
Chicken	28.0	22.0	21.6	28.4
Rat	28.6	21.4	20.5	28.4
Human	29.3	20.7	20.0	30.0
Grasshopper	29.3	20.5	20.7	29.3
Sea urchin	32.8	17.7	17.3	32.1
Wheat	27.3	22.7	22.8	27.1
Yeast	31.3	18.7	17.1	32.9
E. coli	24.7	26.0	25.7	23.6

DNA Composition Anomaly-Possibly, one can tell differences in various animals simply by looking at the mixtures of the 4 ingredients that make up DNA. Please notice from the chart [preceding] humans are very similar to grasshoppers and substantially different than a Chicken or Rat. Oops! We will have to look for a better way, unless you have an exoskeleton I don't know about.

DNA of Homo Sapien Cognatus in North America, using Haplotyping short hand, shows a major mutation of T without the T2 mutation. This means Cognatus came from the Middle East around 6000 years ago. Interestingly, the Cherokee nation has a similar DNA mutation lineage. Yes, Cognatus is sometimes called

Sasquatch, but new evidence is making us look at this strangeness with a little more intensity.

Only Two Races of People- Complex mapping of the human genome shows, conclusively, by Pairwise nucleotide separation and by number of differences in DNA strings; there are only 2 specific races of humans living today. Possibly, this has been hidden to protect one group of people and to amplify the lies. The data is shown below.

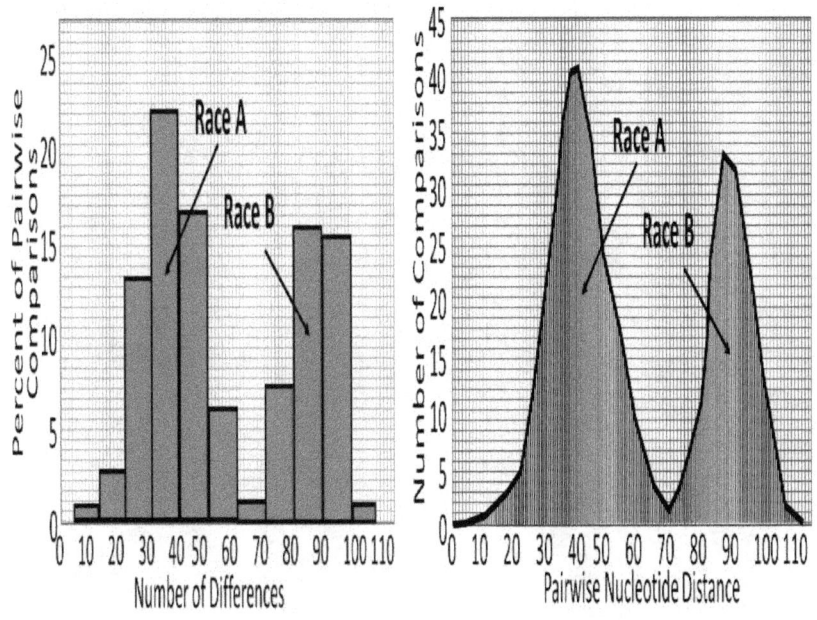

These and many more odd things await your examination. They are generally called anomalies so bogus theories can be acclaimed as factual and our children are placed in jeopardy.

As this book is topical, it is not generally laid out in chronological sequence, but dates are provided to help position anomalies as you go along. We'll start with a little background in the new science of DNA testing and haplotyping. Once we understand a little more about the oddness of Haplotyping and tracing of haplotypes [Haplogrouping], we will have the initial capabilities to shed light on DNA anomalies.

DNA Mutation Tracking

I need to give you a quick overview about DNA and the weirdness of Mitochondria. DNA is a secret part of you that tells where you came from. Instead of saying race, we now say haplotype. Simply put, a haplotype is a colony of individuals that have mutated the same way. Yes, I know that is what we call "race" but English is changing. It seems that major changes in characterizations are, sort of, recorded in the DNA structure such that one can look at a sequence of parts called Alleles that are stuck onto what can be referred to as a "Genetic Segment" and from them determine your ancestral "Haplotype". People with similar Haplotypes are similar because their DNA is somewhat similar. This Haplotype thing not only clusters people by characterization but also tells us when the "grouping" was formed. Therefore, one can determine differences of people groups, societies, continents and how these groups moved from one place to another. I know you have seen DNA used in a courtroom to tell if someone killed another if the killer left a hair in the room, but I'm talking about using this stuff the trace humans back through time to their origins. Here is where the sex chromosome and Mitochondria come in.

Ancestry

Two different Haplotypes can be used to determine ancestry most easily. One is found in a Y-chromosome [Y-DNA] located in the nucleus of every cell in your body and the other is from mitochondrial DNA [MtDNA] that floats around the cell outside of the nucleus. These Haplotypes have different designations. Haplotypes pertain to deep ancestral origins dating back many thousands of years. According to "most" research, Y-DNA is passed solely from father to son, while MtDNA, almost exclusively, is passed down the "maternal line" to both sexes.

Mitochondria DNA

This whole mitochondrion thing should be getting you confused. If not, let me enlighten and confuse. When someone talks about Mitochondrial DNA, he is NOT talking about the DNA that makes you who you are. Instead, people have a **second set of DNA inside themselves from aliens** called mitochondria. Mitochondria are small "quasi-animals" that lie in the cytoplasm of your cells. It is believed these were once completely separate bacteria, but over time, they became "part" of the cells and help supply energy to the nucleus [where the 23 pairs of "real" Chromosomes are hiding as shown below].

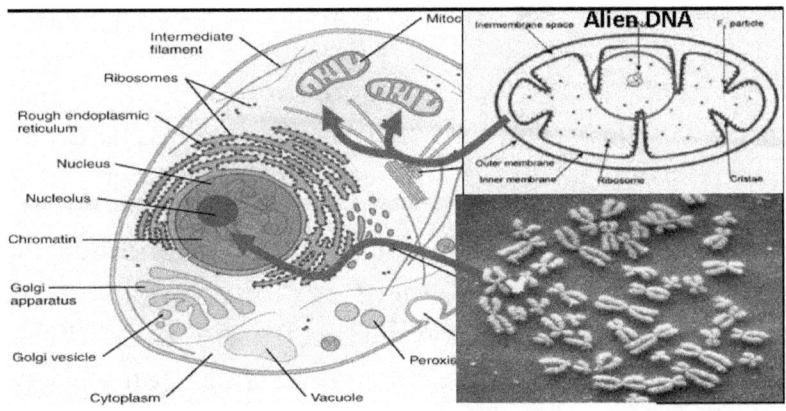

To show it is an alien, the mitochondria has its own DNA different than that of the cell that is copied from Mitochondria in the mother [most of the time]. If the mother had gone through 50 mutations since the animal or person became who she was when she carried a new baby, the mitochondrial DNA that was passed on would be an exact duplicate and carry the information of those 50 changes or mutations. All one must do is decode those 50 codes and every change, ancestor, and mutation would be known. How simple is that? To make it easier, some scientists calculated that there was supposed to be 7.5 mutations every million years. **Unfortunately,** for those trying to make this simple; mutations are very randomized and clumped together during various cosmic events. A number of estimates have the mutation rate at a MUCH higher number, making all the timing much more compressed than previously determined. Later, we will discuss a more correct timeline than the one you have been forced to use from those who are less than sincere or have an agenda to make you believe something that never happened. Let me give you an example.

You probably were convinced that 212 million years ago, the Pacific Ocean was scooped out of the earth. New findings strongly suggest this catastrophic event actually happened about 300 thousand years ago.

This false expansion has twisted things around to suit the proving of various theories that lacked any reasonableness factor. I will briefly go over the new timing determinants in case you have not been shown the massive errors being forced on your and you children to make you believe something that is not there. As a note: this is not what has been called a "Creationist View" of reality where the entire earth and all its calamities occurred in only 6000 years. That view has just as many issues as the old Nuclear decay time-base methods you were forced to use in school. With that, let me get back on track.

Why are Mitochondria from Females?

I suppose you are wondering just how mitochondria DNA only come from women, but the idea is that [most of the time] when sperm enters an egg, its tail, holding the sperm's mitochondria doesn't make it. Either the sperm-tail is simply amputated or when it enters, the tail dissolves so the new entity HAS to use the mother's mitochondria. Unfortunately, for those trying to make this simple, male tails are not always consumed, so a few mitochondria are passed by the male. The image below shows sperm trying to get into an ovum. The picture has been blown up a little for you can see the little rascals.

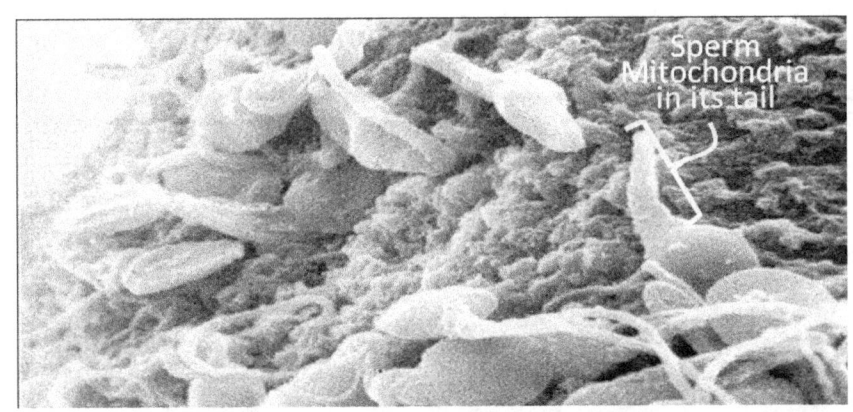

Y-DNA

Similar to the Mitochondrial DNA, Human Y-chromosomes hold mutation details of a male and most of the Y-chromosomes are not even recombined during association so the record of mutations is also held by these Y-Chromosomes. An example might be that if there were 2 generations of identical Y-Chromosome males followed by a mutation. Two sections showing the previous structure and one section showing the mutation would be recorded on the Y-chromosome so one could tell number of generations, number of mutations, and the general time period for each.

Tale of the First Man and First Woman

For any of this to have major truth, there must be a viable community or survival would not have been possible. That being said, the first man and first woman have not been determined to have been near each other when the first inception happened. By common thought, the mitochondrial EVE [first female human] is much older than the Y-DNA Adam [first male human]. That just makes no sense to me at all! I'm thinking Eve was with someone else or the timing is flawed in some way! The other thing that makes absolutely no sense whatsoever is the following.

Human Type	where	First [X000] Yr.Ago	New	100	90	80	70	60	50	40	30	20	10
									[X000] Yr.Ago				
Sahelanthropus	C.Africa	6,000	100	x	x	x							
Australopithicus	S.Africa	3000	95	x	x	x							
Paranthropus	C.Africa	2700	94	x	x	x							
Naledi	S.Africa	2500	90		x	x	x						
Ergaster	S.Africa	2000	88		x	x	x						
Georgicus	Georgia	1800	85		x	x	x						
Heildelberg	Spain	1500	84		x	x	x						
Antecessor	Spain	1200	82		x	x	x						
Habilis	S.Africa	1000	80			x	x	x					
Erectus	China	900	78			x	x	x					
Denisovan	SE Asia	400	65				x	x	x				
Idalto	Ethiopia	160	60					x	x	x			
Neanderthal	Germany	100	55					x	x	x			
Floreaiensis	Indonesia	95	50						x	x	x		
Grimaldi	Italy	90	45						x	x	x		
Boskop	S.Africa	60	40							x	x	x	
CroMagnon	Israel	40	35							x	x	x	x
Modern Man	various	10	10										x
Y/mtDNA	A/L		100		x								
y/mtDNA	B,D/L1, L2,L4		50							x			
T/mtDNA	F/N		40								x		
mtDNA	L4,C,D,E,G,Q,N,Z		12										x
mtDNA	A,B,F,H,I,J,K,P,R,T,U,V,W,X		6										x
Y-DNA	C,H,L,T,G,K		12										x
Y-DNA	I,J,P,R,M,N,O,S,Q		6										x

The chart shows the varieties of hominids found so far and the general dating of each, I put the nuclear decay dating and the newer dating for completeness. The bottom part of the chart shows all the major mutations that define races and differences, written in the Haplotype shorthand. Please note that when there were many differences there were almost no mutations and when there were almost no variations, almost all the mutations found have occurred and then in the last 5 thousand years there have been none. We will get into the "whys" later, but this seemed like a good way to get you thinking.

Ancient Scientists

For some, the next statement may sound strange, but from a huge amount of evidence, including texts and much physical evidence, we can be almost certain the Cenozoic scientists were breeding, manipulating, and changing animals, and people. They, probably, also did like we do today, including modification of corn DNA so a corn worm won't eat it, or modifying pigs so they get so fat they can barely walk. They probably even modified cows like we do today so that they make human organs scientists can retrieve. I'm not saying these people were any worse than us, but they carried out a substantial amount of animal modifications. We know, for instance that they "recreated" Tyrannosaurus Rex, Hadrosaurs and a few other dinosaurs during the Pleistocene and the animals were here during one of the worst wars witnessed by mankind just before the Pleistocene Extinction.

Nuclear War Evidence

Without a doubt, this war turned nuclear using the materials from the Oklo nuclear Plants or others we have not found as of yet. We have massive amounts of evidence of the war turning nuclear. After the war; radioactive, non-fossilized dinosaur soft tissues were left behind and are being recovered every day. The historians and scientists alike simply called the time the "Young Dryas" and said the massive signs of radiation were anomalous; the radio-active, unfossilized Dinosaur remains were anomalies; the massive mutation of human DNA was an anomaly; the hundreds of ancient texts describing our ancient issues were all anomalies; the human footprints walking with dinosaurs were somehow faked; the extremely ancient 16 nuclear processing plants with missing material were "natural"; and on and on we could go as the cover-up continues to this day.

The war didn't matter much to the winners and losers. This is because well over ½ million meteors struck the earth not long after the war ended. The Earth then shifted it on its axis and almost no one survived. While this was happening, those who did survive had been changed as more major mutations of humans occurred at this time, by far, when compared to the previous 90 thousand years. The reason you are thinking this can't be true is there has been an effort to erase details of our history to "protect" us. This book will provide you the information you need to have a much better understanding of human development. We are going to do this by looking at DNA and physical evidence. Let's look at regenerated dinosaur evidence.

Regeneration of Dinosaurs

The recreated animals we know about the most are the reconstructions of Dinosaurs. This all happened about 20 thousand years ago. By this time, the original dinosaurs were all gone, but some DNA still was viable so scientists being scientists, remade the things. Now we are finding seemingly impossible, unfossilized remains. I'm not saying this got our creator mad, but these regenerated dinosaurs were definitely placed in the "unclean" pile.

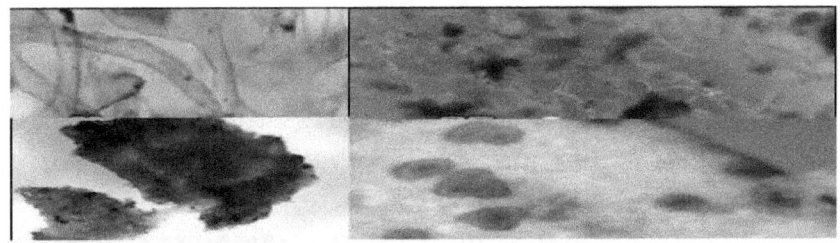

The images above are various blood cells and similar stretchy things and below we see even more of the almost alive parts of ligaments and similar tissue of Dinosaurs.

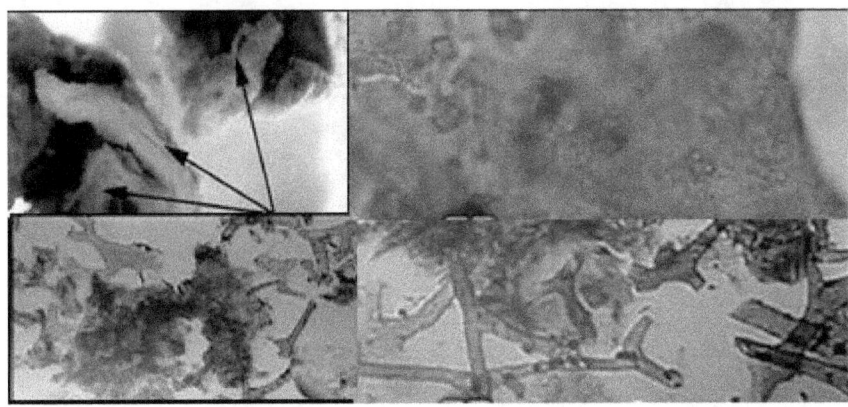

Hundreds of samples found recently show that during the Pleistocene, dinosaurs were recreated and walked with humans again. The biochemists and genetic DNA splicers didn't care about the new dinosaurs not being able to survive well, as their own weight would have almost crushed them. They made them anyway. It is believed the T-Rex constructs would have been hand fed as they certainly could not have run to catch food during the Pleistocene. During this age the earth spin had slowed so everything was heavier than it would have been during the Mesozoic Era. This included the massive head of this out of place dinosaur. Pterodactyls also could fly during the Mesozoic, but today, that feat would not be possible.

In 2012, researchers analyzed multiple dinosaur bone samples from Texas, Alaska, Colorado, and Montana. C-14 dating revealed that they are less than 30,000 years old. The list of dinosaurs found, that were remade during the Pleistocene, keeps growing as with all reptiles, they were considered to be unclean and something else was going on. A graphic following shows the main ones we know about to date. The Pleistocene designers were having so much fun.

More Evidence

Some of these ancient people must have seen Jurassic Park and started finding dinosaur DNA which they turned into a new group of dinosaurs that lived during the Tertiary and Pleistocene Ages. Around the world and in our Bible, we read about unclean or "abominable animals". The reason most of the animals of this time were considered abominations was a massive thrust for geneticists to modify as many animals as possible. I know that sounds weird, but we need to follow the evidence instead of trying to stay comfortable. Besides our Bible, we find out from other Jewish texts the same thing as scientific research in Genetics and Engineering brings us to a dangerous level. I know this still sounds like some bizarre fiction, but bear with me for a minute as we review another tiny portion of the Judeo-Christian texts concerning this seemingly erroneous fact pushed in the Bible. I don't want to bog you down with hundreds of documents, but here are a few to give you confidence about what I'm saying.

Bible-Jubilees 5:3--*and all flesh <u>corrupted its way</u>, alike men and cattle and beasts and birds and everything that walks the Earth all corrupted their ways and their orders.* [The only way that animals corrupted their way was that they were genetically manipulated and just weren't the same animals.]

Jubilees 7:24- *Afterwards they <u>sinned against beasts</u> and birds and everything that moves or walks upon the Earth.* [There are two ways to sin against beasts- sex and genetic manipulation. God didn't like either.]

Jewish Essene-Generations of Adam 6:1-10- AMONG *our little ones - <u>Timnor's sister Ammah</u> was also blessed with understanding, for she investigated the nature of life, unlocking the mysteries of life itself. -- Ammah was not one whit behind her husband in creating wickedness, for <u>she manipulated the very fountain of life, until she had created new forms of beings dedicated to evil</u> and the destruction of mankind.*

Essene-Generations of Adam 8:4-20 *Timnor and Ammah practiced every abomination. Tranter learned the ways of his mother Ammah and he did manipulate the very nature of man and beast <u>to create new forms which God had not ordained.</u>*

Essene-Book of Giants- *- they knew the secrets of heaven and sin was great in the Earth. They made* *mistakes and they killed many animals and people. They selected two hundred donkeys, two hundred asses, two hundred rams of the flock, two hundred goats, two hundred other beasts of the field. From every animal, and from every type of human was taken its seed for mixed sex. After a time, they defiled the animals and people and begot giants, monsters, and dragons.*

Essene-Book of Secrets- *Those who would penetrate the origins of knowledge, along with those who hold fast to the wonderful mysteries of life. -With this I beseech your attention. All of the secrets of manipulating life were known but they [the ancient humans] did not know the secret of the way things are nor did they understand the things of old. Belial who modified creation, a thing that ought never to be done again, except by the command of his Maker. You have not become wise in understanding my secrets of life and the earth; for you have not properly understood the origin of Wisdom.*

Bible-Jasher 4:16-18- *and the sons of men in those days took from the cattle of the Earth, the beasts of the field and the fowls of the air, and <u>taught the mixture of animals of one species with the other, in order therewith to provoke the Lord</u>; and God saw the whole Earth and it was corrupt, for all flesh had corrupted its ways upon Earth, all men and <u>all animals</u>. ---And after this <u>they sinned against the beasts and birds</u>, and all that moves and walks on the earth.*

Just keep all this in the back of your mind as we look at DNA and DNA mutations that will help us understand that most things described as anomaly are described that way simply to protect an inappropriate theory that some group feels comfortable with.

Spontaneous evolution can only reduce complexity of an organism because of the law of Entropy.

The theory of survival of the fittest has not shown to work and many of the species we have today are far less capable of long-term survival.

The theory of God only controlling Creational development of species doesn't work in that massive numbers of mutational mistakes show the developers of some of the species did not realize implications of some of the DNA changes. Luckily, we have DNA to help us find answers, but what do you known about Mitochondria?

Mitochondria Anomaly

Have you ever wondered how these "I-can-find-your-ancestors" companies work their magic? Part of it is done with an alien. I don't know if you know it or not but there is a tiny alien living inside each of your cells called mitochondria, that I brought up briefly before. At some past time, it may have been harmful to us, but now our cells sort of use mitochondria as a useful entity and there are a lot of these little quasi animals. The tail of a sperm, for instance, contains many mitochondria and they run in a spiral-like form along the length of the tail. In heart muscle cells, about 40% of the cytoplasmic space is taken up by mitochondria. To give you an idea what this might mean; in liver cells, the cytoplasm percentage is about 20-25% with 1000 to 2000 mitochondria per cell. If we assume there are about 500 times as many mitochondria as cells as "real" cells in our bodies, we might think they are the main entity and we are some symbiotic attachment. Forget I said the scary stuff as Mitochondria convert energy. The way this is done seems to be good and bad for us as Mitochondria can kill our cells.

Kill and Poop Anomaly

During its normal life, the mitochondrion releases a chemical called cytochrome c, and this can trigger programmed cell death or apoptosis. I know that sounds bad, but sometimes you want some of your cells to die. Besides this death control, mitochondria are also thought to influence, which eggs in a woman should be released during ovulation and which should be destroyed by releasing that cytochrome C stuff. Mitochondria's main job is energy conversion. They eat the sugars and expel/poop out "reactive oxygen species" (ROSs), including some we call 'free radicals'. Oddly, ROSs can damage DNA and mutate Mitochondria so Mitochondria mutates as much as 10 times as often as the real DNA. Sometimes these mutations cause damage in areas needing high levels of energy like areas of the brain, muscles, central nervous system and the eye. Some of the effects are Parkinson's or Alzheimer's disease. If the Mitochondria mutates quickly, the host body ages more rapidly. With all that, no one really knows how the Mitochondria are controlled.

For genetic scientists, they are interested in the Mitochondria's DNA as it is different than your real DNA. Mitochondrial DNA [mtDNA] mutates differently than does our "own" DNA. Here is the really neat part--- SPERM. Sperm, like all the other cells in your body have DNA in its nucleus and another whole set of mitochondria DNA in its tail. As it swims to be the first into an ovum, if successful, the head easily penetrates the egg wall, but then something bad happens, the egg's outer shell becomes hardened and no other sperm can enter. This happens so quickly that the Sperm tail is amputated and that is where its mitochondrial DNA is, I mean was, located. Let's have a moment of silence for the sperm tail. The images show the first sperm entering and then the swarm unable to penetrate. Somewhere in the mix is a tail floating by.

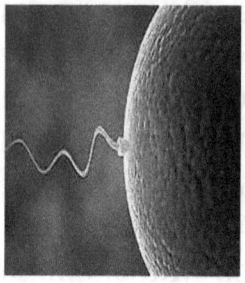 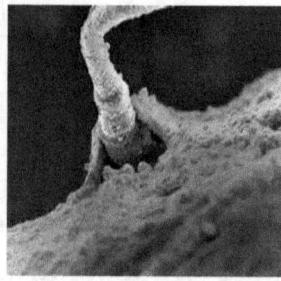 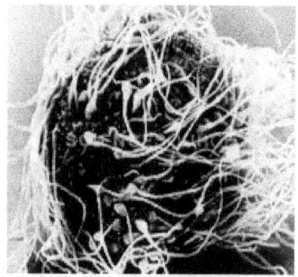

Why is This Important?

The first reason this is important is that a sperm doesn't make a baby. Instead, only the Sperm-head or nucleus is involved. Because we keep losing sperm mitochondrial DNA, a new baby only carries mtDNA from the mother. We can see mutations of both DNAs and if we are testing the mitochondrial DNA, we know it comes from ONLY female ancestors while the "normal" body DNA sometimes called "nuclear DNA" only traces family generations [male and/or female]. Many times, the nuclear DNA is obtained from the "Y-Sex" Chromosome in a cell. A description of this detail is shown next.

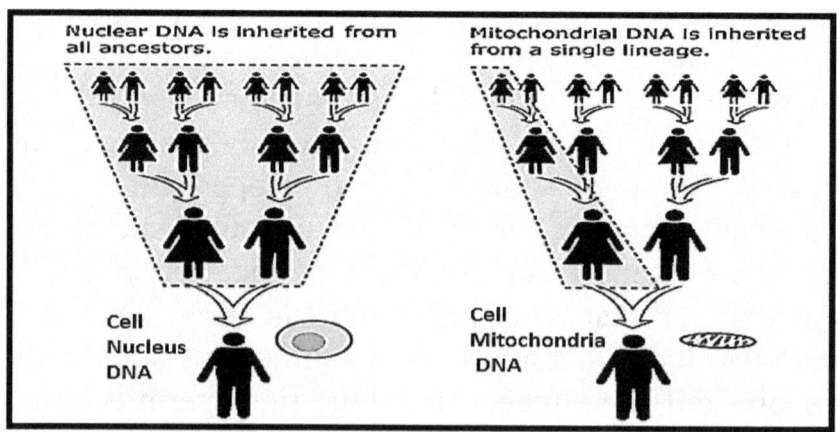

Mutation Memory Anomaly

DNA is interesting in that it has a memory of mutation. Each time it mutates, the "scar" of that mutation is placed in the chromosomal string as a specific change in one of the sugars making it up. As more and more scars are built up a record of when each mutation occurred and how serious the mutation was. We will mostly look at the really serious body changing mutations to find out more about human DNA truths. As you can imagine, there are many more major mutations in the "Nuclear" side as the record passes through more individuals, but we find that family modifications can be quantified into smaller genetic mutation so we actually have a smaller set to review. This will make sense as we go along. The most serious mutation is depicted as a capitol letter followed by a less serious one as a number and then comes a lower-class letter followed by another number. Once the coding gets too complicated, additional codes were made for "groups". Let me give you an example. [R1b] this is the Y-Chromosome mutation associated with Europeans. Depending on where on the Y-chromosome this set of mutations is located tells a researcher when a person came in contact with a European, how many times and the timing of each of the various mutational groups interacted with the test subject. By simply counting the number of [R1b] groups one can tell what percentage a person is with respect to northern Europeanism and the mtDNA mutations will further describe locations of encounter. As an example, the mtDNA mutation "V" indicates a Scandinavian heritage. Therefore an [$R1b_m$:V_f] mutation signature [Y-DNA followed by mtDNA] describes a more pure European Scandinavian than an $[R1a_m:V_f]$ mutation. This whole Bio-engineering science is generally termed Haplotyping and determination of how a group comes into being is called Haplogrouping. This is incomplete when it comes to Homo-Erectus because DNA eventually deteriorates. This doesn't happen when you die as your DNA after death is the same as DNA when you were alive but after 50 thousand years or so, many of the DNA sugars are broken down and the history becomes incomplete.

Haplotyping

I thought I had better give you information on how genetics engineers and heredity tracing engineers go about reviewing our past because you can't find out on these lawyer TV shows.

The problem is that many well-regarded theories could be shattered if anyone knew about DNA. Oops! While many know about DNA sequences, there are just as many reluctant to throw away "quasi-proven fallacies".

Ancient texts and bones, for instance, confirm there was a large number of races before the end of the Pleistocene event 10 thousand years ago, but so does our DNA. As we just discussed, if you are a male, you will have two Haplotypes, meaning Mitochondrial DNA (female line) Haplotype and the Y, sex chromosome, DNA (male line) Haplotype. If you are female, you have the same, except your sex chromosome DNA comes from the "X" chromosome. Interestingly you can tell the sequence of the mutation by where it is and what was before and after a specific mutation [Haplotype]. By looking at similar mutations, we can determine ancestry and even some detail concerning when particular races were formed.

Categorizing Races

Generally, some pairs of Haplotypes are from various races as indicated in the following list. The Haplotypes are coded. The first letter is the oldest and most important mutation. This is followed by a number, then a letter etc. to look for subgroups. The listing following shows the Y-DNA [m], then mtDNA [f], and finally the general grouping of people associated with these special DNA mutations.

Negroid Races
- *Am:(L0)f --Southern Africa*
- *Bm:(L1/L4)f --Middle Africa*
- *Em:(L2/L3)f --Africa wide*

Mongoloid Races
- *(D/O/N)m:(C/Z/D/G/A/B/F)f --East Asia, Siberia*
- *(K/M)m:(B/P/N/Q)f -- Oceania*

Nordic Races
- *(R/I/T/J/E)m:(R0/H/V/J/T/U/K)f--Mediterranean Areas*
- *(Q/C3)m:(A/X/C/D)f--Easternmost Siberia*

Categorizing by Mitochondrial DNA

Here is another grouping that may also be useful by mtDNA mutation groupings. These people can look like cousins but have different mtDNA mutations as indicated.

Amerindian Races
- *Native Alaskan, Inuit: A/D-- Mexico: A/B*
- *North America: A/B/C/D/X /T [Cherokee]*
- *Latin America: A/B/C/D*

Nordic Races
- *Western Europe: H/V/I/J/U/K/T/W/X*

Jewish and Armenian Races
- *Middle East, North Africa: I/J/U/K/T/W/X*

Negroid Races [L]
- *East Africa: L3/L4/M*
- *West Africa: L1/L2*
- *South Africa: L0/L1/L2*

Mongoloid Races
- *South Asia: N/M/U/B/F --Melanesia: P/Q/B*
- *Australia: N/P-- East Asia: N/M/B*
- *Eastern Russia: L/Y/G/A/C/D-- Russia: Z*

Haplogrouping

Besides looking for heritage, we need to determine how people got where they are. That study is called Haplogrouping.

Haplogrouping is the study of placement of the haplotype mutation in a chromosome sequence. The sequence tells the trained professional the lineage of people by what mutation was locked in at a particular time. Say for instance, the "A "male haplotype was followed by an "F" male haplotype, followed by an "F" male haplotype. This would tell the researcher the individual roamed from Africa to the Middle East and finally to southwest Asia. With this technology, one can trace family lines, heritage, pureness, and where people were when each mutation addition was added. From this testing, it has been scientifically determined that about 11 thousand years ago almost ½ of the "major" mutations of modern humans occurred and almost all the remaining ½ mutated about 5500 years ago. If you are confused enough, let's continue. I think seeing some maps will help. Any haplotype and haplogrouping in this book we be exclusively looking at major mutations of man.

Y-DNA Haplogroups

As I said, by testing large groups, one can map out where individual groups came from, and how the lineage took control of various places around the globe. The following map shows a generalization of this type of Haplotype flow-map. Following the map are general descriptions of the various mutation grouping and a time-period for each event. The relative timing of the events is similar to known tracking, but I have compressed the timing so we can be closer to the ballpark. While we cannot get an exact time, we can determine what mutation comes first so we can adapt the sequence to known events. Hopefully, from my previous discussions it is known that while it is the best thing we have, Haplotyping is not an exact science. There have been so many intermarriages and so many thousands of years, who came first, the Hamite or the Gaelic, can make it difficult to track entities and attempt the generalization of how a person got to be who he is today. Also, note that I confined the African mutations to Africa. While there certainly were extrusions, especially by the "E" grouping, these would not happen for a while.

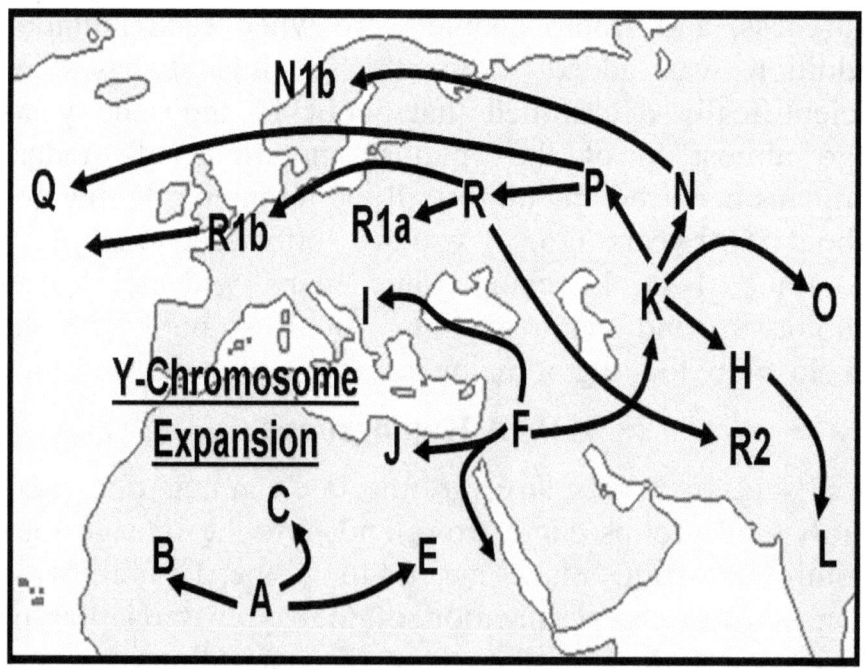

One can define major mutation or combination points by a letter and number identification. The first letter is the most significant ancestral point. It denotes MAJOR mutation points so we can time them pretty well. The number following indicates a single event mutation/modification and a second letter identifies an additional subset or combination of groups coming together. Please note the trail of "major mutation points" is characterized by general characteristics of the groups living in those locations and how they SEEM to flow. For the example above, the string "F to K to P to R" shows a particular timeline. This is believed to be the chain of ancestry for Europeans. Of particular importance is the Haplotyping known as R1b which is the base grouping of Northern Europeans and Eastern North Americans. Sometimes names are given to these groups to make identification easier. The dates are approximations for reference and I have added the names to help classify and look for anomaly.

Y-Chromosome Haplotypes

The study of DNA is a fairly new science, but it is rapidly getting more and more detailed as Haplotyping of DNA mutation is used to find out the ancestry of just about anyone today. A study dating the age of more than 1 million single-letter mutations in the human DNA code revealed that <u>most of these mutations are of recent origin</u>.

*According to one study, over **86 percent** of <u>the harmful single nucleotide mutations</u> arose between 5 and 12 thousand years ago. Oddly, since then there have been very few major mutations.*

Overall, the researchers in this study now believe that about 81 percent of the single-nucleotide variants in the European sampled and <u>58 percent</u> in the African DNA sampled arose in the past 5 or 6 thousand years.

In the African samples, a large number of the single nucleotide mutations appeared before Cro-Magnon [40 thousand years ago], so those would have had more Homo-Erectus ancestry, but among Europeans and the rest of the world, many mutations occurred between 10 and 12 thousand years ago.

By looking at sequential mutations recorded in "Y"-sex chromosomes, the following list of races around the world whose beginnings have been fixed and decoded. Please notice most changes occurred either 10 to 12 thousand years ago, during the time of the great Pleistocene Wars; or 5 to 6 thousand years ago as I stated earlier which will coincide with a second Great War called the Kurukshetra War, Bharata War, or the Tower of Babel War. I know these wars have been eliminated form history books, but that does not mean they weren't horrible and defined who we are today.

Mutation indicator	Time [x1000 yr]	Races of Men caused by mutation
Y	100	First Y-DNA
B	50	Sapien Sapien
F	40	Cro Magnon
C	20	Negroid
E	12	Nubian
G	12	Armenian
I	12	Greek
K	10	Asian
J	10	Canaanite
R	10	Scythian
H	10	Afghan
P	6	Proto Ameridian
R2	6	Aryan
N	6	Russian
O	6	Oriental
L	5	Dravidian/Indian
R1a	5	Slavic
R1b	5	Gaelic
N	5	Scandinavian

The elimination of major segments of our history simply means historians, many times, are not honest or they want to believe something so badly, they erase from their minds actual truth and replace it with vain truth where desire out-trumps logic. Hopefully, we can remedy some of that. The beginnings of the Nubians, Armenians, Greek, Asian, Canaanites, Scythians, and Afghans all are traced during a horrible War just before the end of the Pleistocene. Please take note that dating Haplotype mutation scars is not an exact science as one must look for which scar was before or after another on extremely tiny objects. You will notice some differences between the various charts throughout this book and just about any on the DNA tracing subject, but generally they show the anomalies presented very consistently. Here is another chart that is slightly different than the previous one.

A= "Y-DNA Erectus" [100 thousand years ago]
B= Sapien [50 thousand years ago]
F= Semitic [40 thousand years ago]
C= Negroid [11 thousand years ago]
E= Nubian [11 thousand years ago]
G=Armenian [11 thousand years ago]
I = Greek [11 thousand years ago]
N= Russian [11 thousand years ago]
O= Oriental [11 thousand years ago]
K=PreAsian [Japheth] [10 thousand years ago]
J=Hamite [10 thousand years ago]
P=Proto-Amerindian [10 thousand years ago]
R =Scythian [10 thousand years ago]
R2= Aryan [6 thousand years ago]
H= Afghan [6 thousand years ago]
L=Dravidian [5 thousand years ago]
R1a=Slavic [5 thousand years ago]
R1b=Gaelic [5 thousand years ago]
N1b= Scandinavian [5 thousand years ago]

Don't worry about all this stuff right now it will make sense as we go along. Also, it should be noted and today some suggest that only the "A" Genome or haplotype descendants came from Africa as the "B" type, possibly, started in a different country and descendants went "INTO" Africa. The mutation points before 10 thousand years ago would be those that were established before the worldwide flood and Pleistocene Extinction 10 thousand years ago. The 6-thousand-year boundary would be from the beginning of the Bharata War and the 5-thousand-year mutation boundary was around its ending [3100 BC]. To give you a little better perspective, here is a similar chart that shows the major offshoots and when each apparently split off. It also fills in some of the missing lines in the preceding Haplotype map I presented. Please don't be upset about the last 2 human type Lizard-men and Ape-men. That story is coming up later as those people did not survive long.

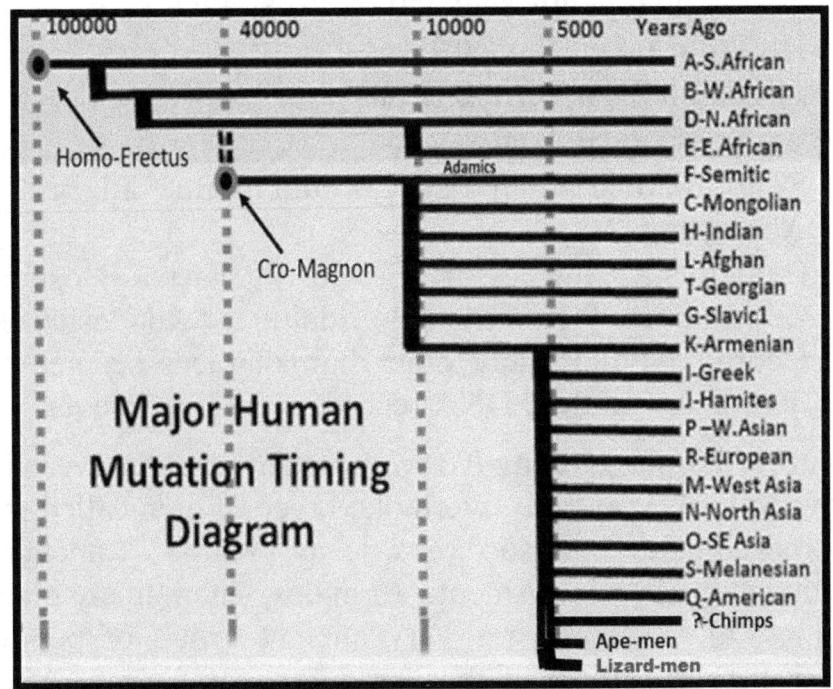

These charts allow us to witness the modifications of the DNA of the Humans that were living so long ago and another DNA sequencing gives us even more data.

10 Major Mutations

As described previously, we find the beginnings of what we see as "races" today as they are mostly initiated at the end of the Pleistocene Age 10 thousand years ago. From the Y-DNA Haplotyping, we find 10 specific massive mutations that will define the major races of people around the world.

- C= true Negroid
- E= Eastern Nubian
- **G=Armenian Who would spawn more of Europe**
- I = Greek [who would become white Nordic groups
- N= Russian
- O= Oriental [Chinese, Pacifica, Australian.]
- K= ProtoAsian [Japheth]
- J= Hamite [Northern African]
- R= Scythian [Balkan]
- P=Proto-Amerindian [That would define all of the Americans.]
- Heidelberg, Antecessor, Erectus, Ergaster, Boskop, Grimaldi, Floresiensis, Idaltu, Rhodesiensis, Georgicus, and many other human races seemed to disappear and their DNA records were lost forever.

Semitic Homogenizing-To find out more truths we will have to look at another overlooked race of men called the Homo-Gigantus. Also known as Homo Capensis, Annunaki, Akamim, Amenti, Archaics, Titan, many other names, all referencing a very ancient group of people who were the rulers of the world for many years. After the Pleistocene extinction, many races had vanished and others had mutated and heavily colonized various areas of the Middle East and Africa. Associated Semitic races are depicted in the following collage left. Lower row mutated 5 thousand years ago. The upper mutations were from the Pleistocene Event.

African Homogenizing-Those who found themselves in Africa after the floodwaters receded, mostly, remained in Africa. While there were mutations, interbreeding makes it difficult to determine various groups.

A general accounting is shown in the preceding right collage. As before, the upper row is from the Pleistocene, while various tribes had a number of similarities. Various tribe images are shown. Besides tracking by Y-chromosome DNA, the Mitochondria things give us another perspective.

Mitochondrial Haplotypes

While you would think tracking mitochondrial DNA would show almost the same expansion and mutation, this is not the case. One reason might be that the mitochondria are more protected so there are more "straight links". Along the left of the map shows the major mtDNA of the Americas. We will have to investigate how the N and M Haplotypes all of a sudden show up in America without secondary changes.

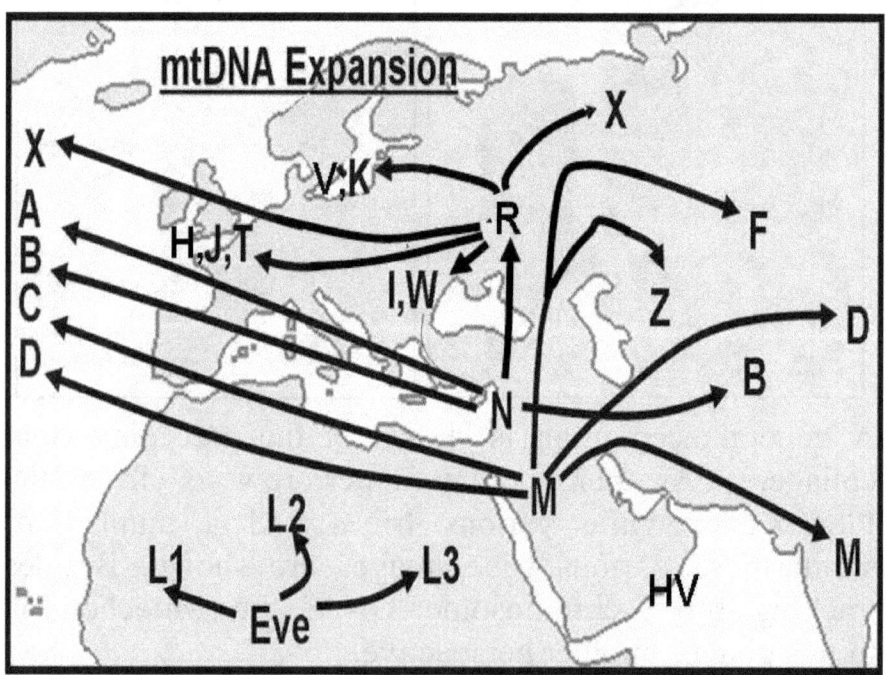

L=Eve=Homo Erectus [100 thousand years ago]
L1= Sapien [50 thousand years ago]
N= Adamic [40 thousand years ago]
M=Arabic [30 thousand years ago]
L2= Negroid [20 thousand years ago]
L3= Nubian [15 thousand years ago]

R = Proto European [11 thousand years ago]
F=Mongol [10 thousand years ago]
Z=Oriental [10 thousand years ago]
X= Proto-N. Amerindian [11 thousand years ago]
A=Adamic-Amerindian [11 thousand years ago]
B=India-Amerindian [11 thousand years ago]
C=Russo-Amerindian [11 thousand years ago]
D=Oriental-Amerindian [11 thousand years ago]

V, K= Scandinavian [6 thousand years ago]
I, W = Greek [6 thousand years ago]
H, J, T2=European [6 thousand years ago]

Like the Y-Chromosome map, there are many variants of these flow maps as well. Again, it should be noted that these are my names rather than ones used by others. Please notice that there are not nearly, as many "Mutations" associated with Mitochondria DNA and the Y-Chromosome so we will concentrate of Y-DNA Haplotyping mostly.

Caution

There is a large group trying to push the E Haplotype [Y-DNA type] into all sorts of interactions, but there is little evidence of that. I will explain some of that as we go along. One of the better-known population distributions concerns the expansion of Ireland and the UK. I will go into that as that population drift affects many and a substantial amount of secondary information can be used to form a more probable lineage of this group as well as others.

Please note that the 5 American MtDNA groups don't make sense.

While MtDNA shows very few mutations [mostly secondary mutations of "L" compared to the Y-DNA, almost overnight the X, A, B, C, and D mtDNA mutations formed on both continents AT THE SAME TIME.

It was as if some of the people were brought over to America in flying machines and they stayed. ----Forget I said that. Maybe all the different mutated people got together one day and made a vow of no-crossbreeding, walked across Asia to that stupid land bridge thing, and ate some polar bears and such as they finally got to the Americas to started breeding again, just like history Textbooks and Teachers are telling our children.

What we see everywhere is a successful attempt at changing history. One area noted is Monogenism where all people started out in Africa. On the surface, it seems stupid and as we peel back layers of inappropriate suppositions, we find that we don't change our mind. It is stupid!

Staying in Africa Anomaly

What if I told you the common ancestor between Africans and Middle Easterners was made up?

As DNA doesn't lie [often] we can get a better understanding of how people got here than listening to someone spout off about all humans coming out of Africa because "Lucy" the Australopithecus Ape-woman was found; the first Homo-Habilis; and then the first Homo-Ergaster was found there. Some might stupidly think the rest of the world was devoid of anything more advanced that a worm, but that is simply not the way these things go.

If you noticed I did put the caveat [often lying] as people can still manipulate things. Such is the case of the first Haplotype family tree of major mutation in the Y-Chromosome. It is shown next.

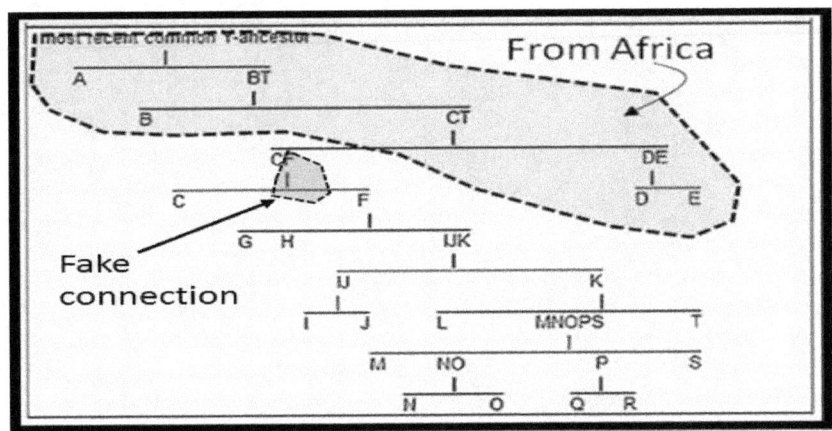

> *The problem is that the "CF" dual mutation never happened as F was completely separate and found in the Middle East while C, along with A, B, and D all were placed in Africa.*

We also now believe C mutation humans did not venture out of Africa, but they seem to have come from Europe; and there is a line to Homo-Neanderthalis very late in development time, so the whole thing is skewed just to make it look like people all came out of Africa. By the way; this chart, and many like it, concentrate on the major mutations. [Capitol Letter]

We will correct this thing as we go along so you can appreciate anomalies that were purposefully ignored when establishing development theories that are still taught in our schools. They call the theory Monogenism [one genus type] Theory.

Monogenism Anomaly

We found Homo-Ergaster in Africa so all humans must have come from them or Evolution is in jeopardy.

Monogenism is the misguided characterization of evolution that is still found in most human race histories. It is the belief that all humans evolved over millions of years and they can be traced back to a single "seed". This is sometimes called anti-Polygenism or Monogenism or the Out-of-Africa "truth". Unfortunately, there is little truth in this idea. It simply allows the theory of uncontrolled evolution to continue without being challenged. For this one the proclamation came that Homo-Ergaster was the first man and the first man and all "Evolved" from him as black people. Over millions of years, some people got lighter, or bluer, or redder, or more yellow and then we had races. I don't know how Oriental Homo Erectus or Java man, red headed Neanderthal, white skinned Cro-Magnon, Heidelberg, and Melanesian Denisovan came along, but you hear, *"Anything can happen if millions of years are used."* The more DNA is inspected the more everyone realizes all this is bunk. Since nuclear decay timing has been debunked as having errors of 5000% and Ice core timing tells us the Cretaceous Extinction happened only 120 thousand years ago, the *"million years thing"* has, pretty much, gone away. It looks more like this.

- Homo Gigantus-180 thousand years ago.
- Cretaceous Extinction-120 thousand years ago.
- Proconsul-Africa-118 thousand years ago
- Gorilla-Africa-115 thousand years ago
- Sahelanthropus- Africa-110 thousand years ago
- Paranthropus- 100 thousand years ago
- Homo Giganticus -SE Asia-98 thousand years ago.
- Homo-Habilis-Africa-95 thousand years ago
- Homo-Naladi-Africa-90 thousand years ago
- Homo-Ergaster-Africa-88 thousand years ago
- Homo-Georgicus-Russia-85 thousand years ago
- Homo Heildelberg-Europe-84 thousand years ago
- Homo-Antecessor-Europe- 82 thousand years ago
- Homo Erectus-SE Asia-78 thousand years ago
- Homo-Rhodesiensis- Africa-78 thousand years ago
- Homo-Denisovan- SE Asia-65 thousand years ago
- Homo-Idaltu -Africa-60 thousand years ago
- Homo-Neanderthalis-Europe- 55 thousand years ago
- Homo-Floresiensis-SE Asia-50 thousand years ago
- Homo-Gerimaldi-Europe-45 thousand years ago
- Homo-Boskop- Africa-40 thousand years ago
- Homo-Cro-Magnon-Mid. East-35 thousand years ago
- Young Dryas radiation period- 11 thousand years ago
- Pleistocene Extinction- 10 thousand years ago

Anomaly after Anomaly

This list is very incomplete as we find more every year and I removed most of the Apes. Every 2 to 5 thousand years another major change in human size, brain size, fingers, walking, teeth, skull, and brain shape. All the while these humans jumped all over the place from SE Asia to the southern tip of Africa, into Germany, and northern Asia. It's no wonder nuclear Decay is still being taught. While there is a logical explanation for the massive numbers of mutations of humans, that is the subject of a different book. What we want to look at here is there are huge numbers of mutations and, all of a sudden, all the mutations were, essentially, wiped clean 10 thousand years ago as if none of these people ever existed. We know something massive happened during the Pleistocene Extinction, but guess what is described in children's text books?

Short Overview of Modern Timing

I can see right now some of you are having trouble with the Cretaceous Extinction happening only 120 thousand years ago. While there are many books on this subject and it is not directly associated with DNA, let me just give you a brief overview of how the timing changed. For a long time, scientists used nuclear decay as the master timing element. If it disagreed with the consensus, the timing was redone until the desired information was obtained. This was possible because nuclear decay is totally random. Anytime a sunspot shows on the sun, the earth is bombarded with billions of neutrinos that travel through the earth. As they travel, they decay nuclear material very quickly. If a volcano erupts the heat decays nuclear material very quickly. If a nuclear event occurs, the nearby nuclear material decays very quickly. After Mt. St. Helens erupted, a single item could get age reading of 500 to 50,000 years. During the last major sunspot activity, nuclear material was tested to have a variable age of between 100 and 500,000 years. During the Pleistocene, massive nuclear events occurred 11 thousand years ago, which would have disrupted timing everywhere. So, how can we determine how old stuff is?

In Antarctica and Greenland, we can core the ice and look for seasonal changes in CO_2 and O_{18} [O18 is important in that it also tracks the abundance of seashell creatures of vast amount of time.]; in the Pacific Ocean, we can track the Hawaiian Hotspot. In the middle of the Atlantic Ocean we can track the Magnetic field alignment. All together we can map extinctions, Earth axis shifts, animal catastrophes, and magnetic shifts. When reviewed together they give us a true picture of earth timing from when the Pacific Ocean was carved out of the Earth until today. The chart following shows the section we are interested in. From top to bottom are the O_{18} [marine isotope percentages], The magnetic field reversals in the Atlantic, the CO_2 concentrations in ice core samples, and the last is the perpendicular to the Hawaiian hotspot track in the Pacific. Please notice they all line up to show extinctions. I placed the names of the eras on the graph to allow you to see the timing more clearly. Even a secondary extinction event during the Cretaceous is depicted.

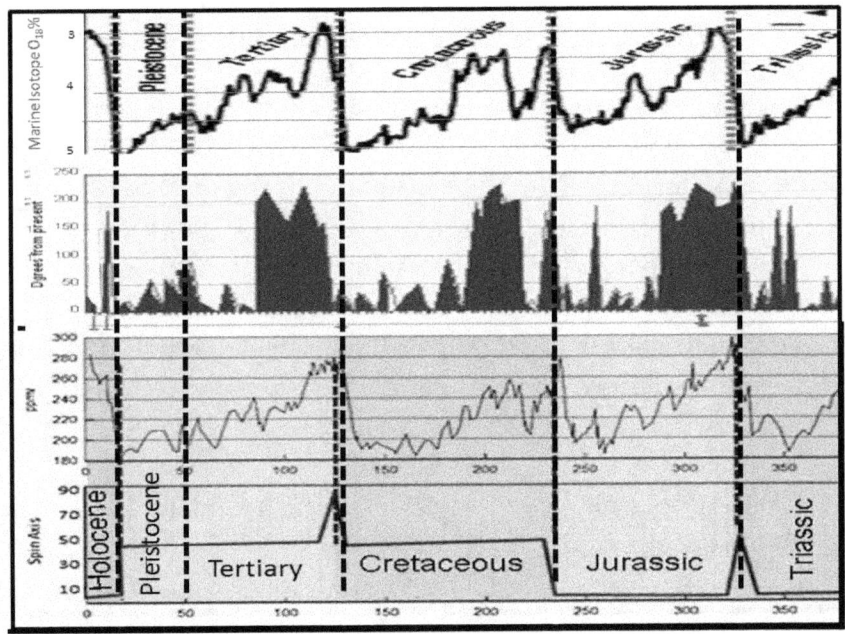

The following geologic timing chart shows the timing previously presented as "nuclear decay-based timing [that you were taught in school] and the more modern characterization of CO_2 density, temperature, O_{18} levels, Paleo-magnetic and Hot spot travel timing. The farther from the 40-thousand-year limit for C14 decay timing, the more variable the nuclear decay timing gets but the newer timing method have no such limitation. There is some reason Cretaceous dinosaurs are now being found "unfossilized and radio-active" as if they were roaming less than 20 thousand years ago. During the Pleistocene Era, scientists reconstructed them just like we saw on Jurassic Park and there was a nuclear war during the Young Dryas period that not only made the remains radioactive, it also mutated humans as we will see.

Standard Geological Timeline		
Era/Period/Epoch	Time (M yrs. ago)	Time (T yrs. ago)
Archaeozoic Period	5000-1500	50,000-3000
Proterozoic Period	1500-545	3000-1000
Cambrian period	550-500	1000-900
Ordovician period	500-440	900-800
Silurian period	440-410	800-700
Devonian period	410-365	700-600
Carboniferous	365-300	600-500
Permian period	300-250	500-400
Triassic period	250-212	400-300
Jurassic period	212-145	300-200
Cretaceous period	145-65	200-100
Tertiary period	65-0.04	100-40
Pleistocene period	0.04-0.01	40-10
Holocene period	0.01-0	10-00

Young Dryas Radiation

Scientists indicate they don't know what happened during the time period called the Young Dryas, but there is a lot of evidence of a massive nuclear war and they simply don't want to describe that. The graph following shows a dramatic change in temperature between about 11 thousand and 10 thousand years ago that quasi-scientists simply call Dryas. While we have been timing the 10-thousand-year event as the end of the Pleistocene; what in the world happened a mere thousand years before?

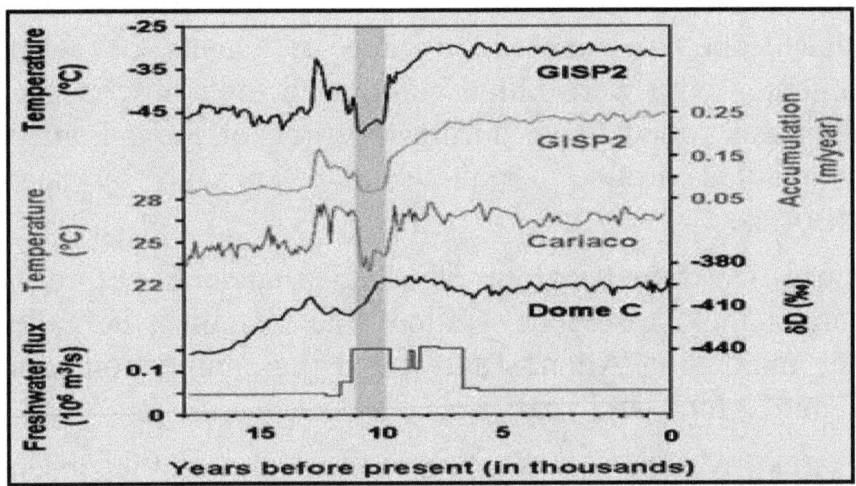

The top two ice core samples are from Greenland and the last 2 are from Antarctica to show this was a worldwide event. While we may not know for sure what exactly happened there are indicators that strongly suggest something terrible happened. Here are the things we know.

Craters- ½ Million meteors came from somewhere around 10 or 12 thousand years ago and fell all across the

Eastern coast of the United States during the Dryas. The massive crater field is called the Carolina Bays.

Temperature-There was a massive drop in temperature around the world about 11 thousand years ago during the Dryas.

History-There are many depictions of a massive worldwide war just before the worldwide flood that ended the Pleistocene. This includes stories from our Bible indicating as many as 1/3 of the population of the world was killed.

Pleistocene War- Some of the depictions of a Pleistocene war include a race of Giants or Homo Gigantus who were killed along with their descendants. This was timed to be during the time of King Lamech who ruled around 15 thousand years ago by some histories.

Plato- Described a group of Islands that sank, before the end of the Pleistocene. Called "the Undalls", he called one major city Atlantis and timed the submersion to be about 12 thousand years ago.

Venus- We are now finding many signs of Pleistocene colonization followed by war on the Planet Venus. These are being timed as "recent" by astronomers. Recently its rotation shifted 90 degrees from all the other planets.

Destruction of Rahab- Clear indications of the destruction of the planet Rahab, the morning Star, can be found in our Bible and other histories.

Weaponry-Descriptions of the use of mighty weapons in a Pleistocene War are found in a number of Bible related works.

DNA- We are told by Haplotype scientists that almost ½ of all major DNA mutation occurred 11 to 12 thousand years ago and most of the other mutations occurred around the time of the Bharata War.

Radioactivity- We will see there was a sharp short-term rise in radioactivity about 11 thousand years ago. We also are finding unfossilized dinosaur remains that are highly radioactive and knowledge of 16 nuclear processing areas in Africa predate the end of the Pleistocene.

Radiation

We know that many animals and plants became extinct at this time. Then an indicator tells us more. *Uranium concentrations in coral jump by almost 300%.* Also, we find marked increases in **nanodiamonds,** magnetic spherules [tiny balls], and carbon spherules at the end of the War with a major increase in charcoal around the middle showing fire and general war conditions. The nanodiamonds indicate heat generated by nuclear explosions. The darker area on the following graph represents the Dryas.

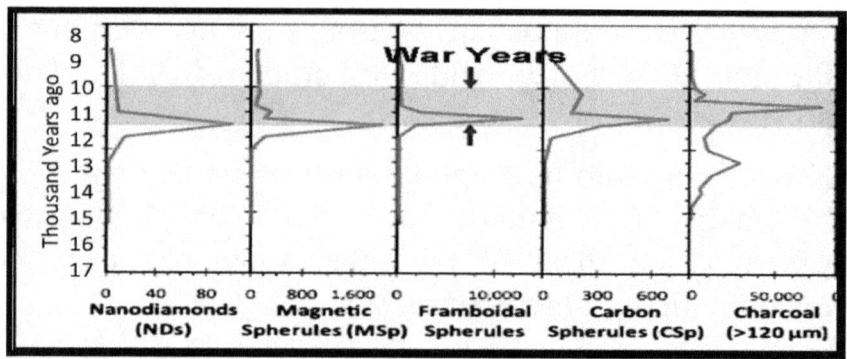

Forensic Science-While there is little doubt that nuclear weapons were used in this ancient war, there is also indications of another type war with spears and bullets. Some scientists started looking for holes and found bullet holes in skulls of Cro-Magnon and earlier people, pointing to the use of high-speed projectile weapons used by the Anak people. Here are just a few examples.

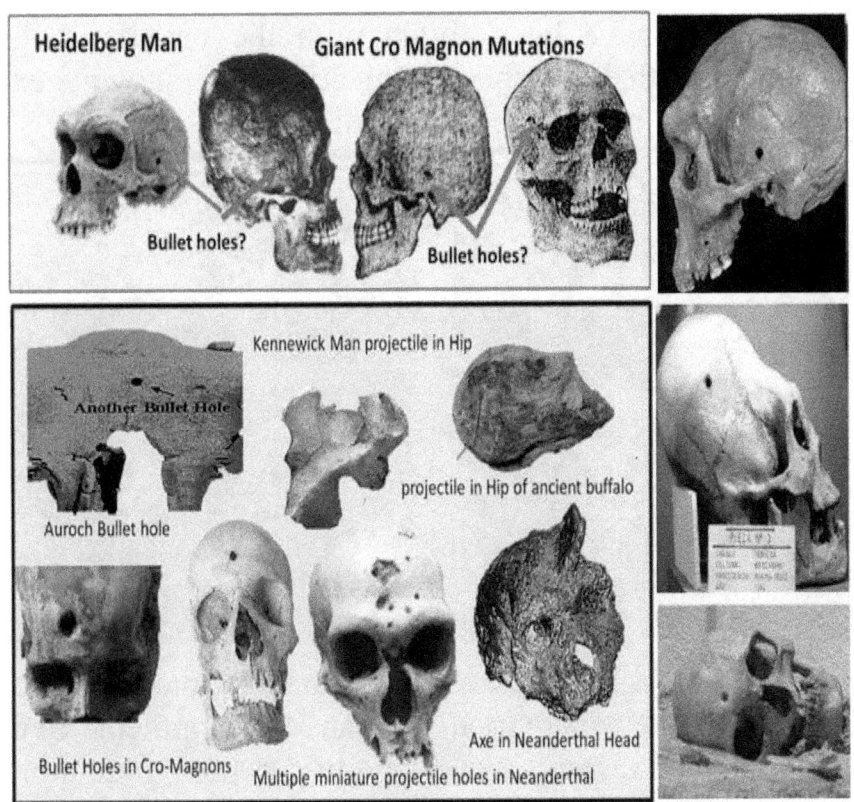

In Russia and extinct Auroch was found with a similar high-speed projectile hole in its forehead which is believed to have been put there during the Pleistocene [See the following collage]. A similar hole was found in the same type animal in Zambia and other signs of hostilities. Most had a tiny hole going in and much larger exit destruction. Please notice something that might be bad. Many of the bullet holes are from point blank range given the entrance hole positions as if done by a contract killer. I wonder if it was some form of mafia.

Other skulls have been found with the peculiar holes associated with high speed projectiles and shown next. Some of these may not be from bullets.

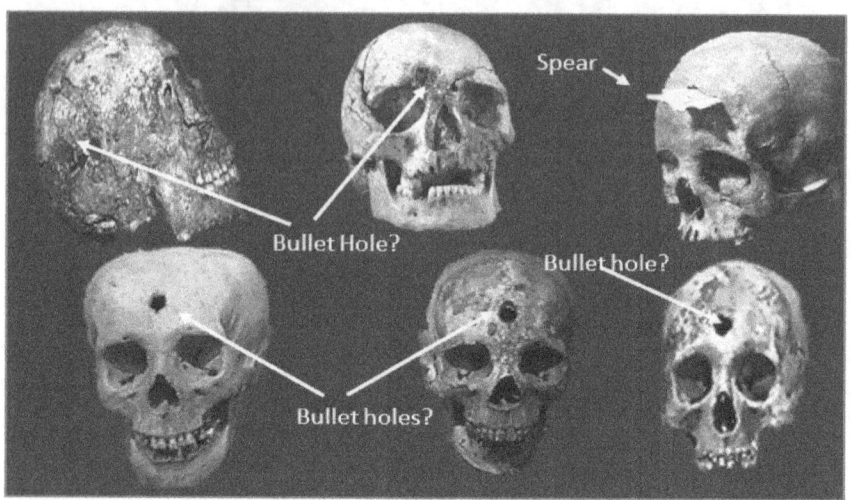

With all that in the back of your mind, let me show you the crazy out of Africa human development map. Initially, DNA testing or Haplotyping seemed to go along with this absurdity and mapping showed how everyone came from homo-Ergaster. Everything flowed nicely. White people "evolved" from the dark-skinned Dravidians of India around the time of the Pleistocene Extinction as an offshoot of what now was the Neanderthal. It should be noted that Neanderthal have never been found in India or Africa and flooded the European countryside during the Pleistocene and then disappeared just like almost all other human races.

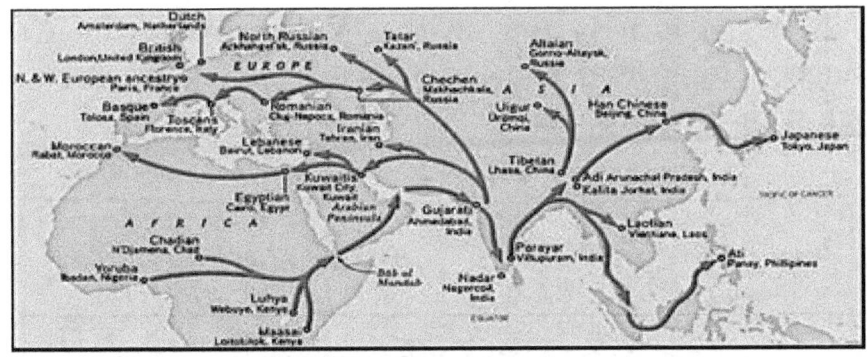

Corrections Please!

Soon it was found that while homo-erectus based people did indeed come out of Africa, but then a brand-new type of man was "spontaneously generated" in the Middle East. The African based people began as black people and were of a much older race than the redheaded Middle Eastern based group. By the time this new human came along, the DNA suggests that the African DNA had spread to Europe and Asia without there being signs in the new DNA humans of the Middle East. Also, we found that Heildelberg and Neanderthal had no ancient African heritage. Rather than the "Out of Africa map, we might see the following with the new data.

In a log scale of time, we find most of the Tertiary humans all died off during or before the Extinction event.

This is not exactly true in that some mixed with Cro Magnon to extend some details of the DNA changes before the modern Holocene Age.

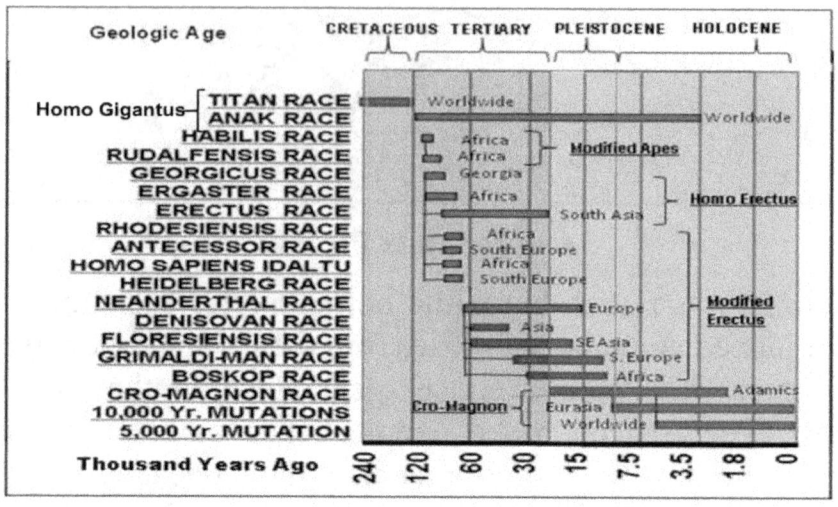

Bibles says Many Survived the Flood

The Judeo-Christian texts tell us, at the land of Nod [located somewhere east of Iran, the descendents of Cain mixed with all the others to become a group called the Gentiles. The Bible says 6 times that *"all the humans on land and not in ships"* died during the worldwide flood associated with the Earth shift and extinction that carried million animal herds of Mammoths from a field filled with flowers to a solid mass of ice in what is now Siberia, quick freezing and killing them all. Some escaped in a wide variety of ships as described in 80 thousand ancient texts about the horrors of the Pleistocene Extinction and flood. The Bible agrees, but some seem to try to read in something else as Noah's were the only pure Cro-magnon blooded humans and all others were gentiles.

Here are 6 attempts Moses had to let someone know other ships were saved.

Genesis 6:17*-I Myself am bringing floodwaters on the earth, to destroy - <u>all flesh</u> - <u>that is on the ground</u> shall die.*

Genesis 7:4*-I will destroy <u>from the face of the earth</u> all living things-* Noah was not part of "all" because he was on a boat.

Genesis 7:21 *And <u>all flesh</u> died <u>that moved on the ground</u>:*

Genesis 7:22 *<u>all that was on the dry land, died</u>.*

Genesis 7:23a *He destroyed all living things which were <u>on the face of the ground</u>*

Genesis 7:23b *All were destroyed <u>from the ground</u>*

Noah and others were on the Earth, but not on dry land so he were saved as the earth axis shifted and the world was flooded.

The following graphic shows the general location of this Nod place and where we found most of the remains of the various Pleistocene humans. The Gentiles made new people by crossbreeding while the Homo-Gigantus people experimented on DNA.

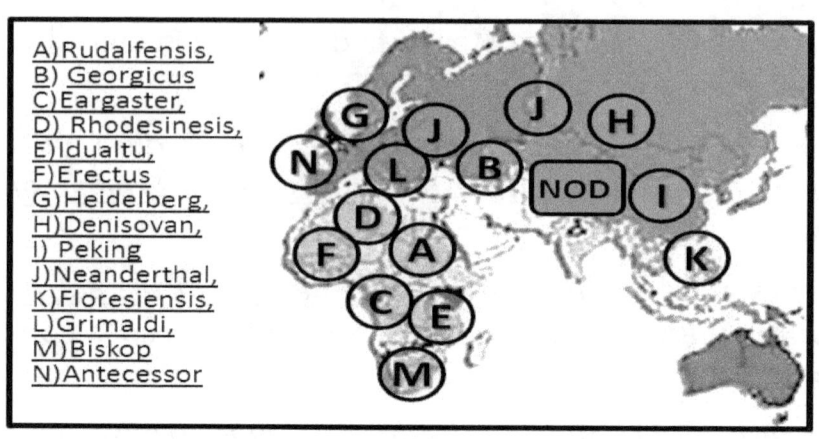

For the average person about 4% of your DNA is known to be from Neanderthal; whereas, about 5% of the genome of Melanesians DNA came from Denisovan. For those still wondering how in the world mixed breed Neanderthal and Denisovan humans could have gotten past the worldwide flood of Noah, please reread the preceding paragraph until you can accept it. Tibetans have a DNA Haplotype mutation that assists with adaptation to low oxygen levels at high altitude and in 2014 it was found that <u>Denisovan had the same mutation</u> showing their close relation. With the massive distances between each of those that seemed to be connected with Denisovan, it is easy to conclude that the Homo Capensis humans [long skulled giants called Anak in Biblical histories] were working to modify and make Neanderthal more rigorous. One group of Homo Capensis must have tried their genetic art in Indonesia. Rather than even suggesting such a thing, it seemed better for consensus scientists and "don't rock the boat" historians to simply bring out the ANOMALY stamp and place all this data in a hidden compartment.

A More Plausible Religious Description

For what it's worth, let me just tell you the Judeo-Christian description of how all this came down. Those reading the Judeo-Christian histories find that Cain, son of the first Cro-Magnon Man, got a huge "family" and waged wars with his father's people. Apparently, the Homo Capensis took some of the Homo-Erectus variants and changed them or they crossbred with the Cro-Magnon family of Cain and Lebuda, his sister/wife. Prior to the Earth shift which caused the Pleistocene Extinction and Worldwide Flood, the various populations of humans were located something like that shown next. These all died out except for the Cro Magnon and the mixing of Cro Magnon either by DNA splicing or human interaction to form the general grouping called Gentile or non-pure Cro-Magnon.

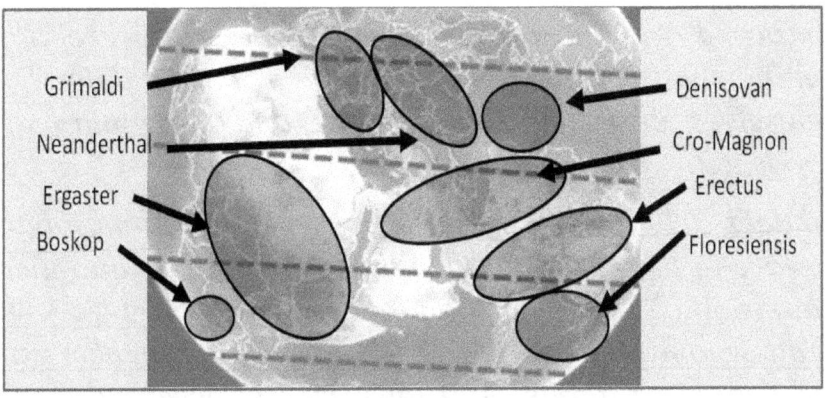

Here are some of the ancient verses describing the development of mixed breed Gentiles during the Pleistocene.

***Codex Junius II**-Then the Anak [Homo Capensis] began to take wives <u>from the tribe of Cain</u>, a cursed folk, and then the sons of men [Homo-Erectus variants and Cro-Magnon] took wives from among that people, the fair* *and winsome daughters of the "Sinful Race" [Cain's offspring].*

Wars broke out and some of the Cro-Magnon descendants had giant offspring after union with Anak [Homo-Gigantus or Capensis] people.

Generations of Adam 7:3- *The people of Timnor and Cain began to come upon our children* [Cro-Magnon or Adam's children] *and stole herds and produce, killing any who sought to prevent them.*

"Cave of Treasures"-*Adam knew his wife again, and she brought forth Seth, like unto Adam, and he became the father of the <u>Giants</u> [Homo Capensis] who lived before the Flood.* [Therefore, he married one of the Homo Capensis.]

Eliezar *-From them [the Homo Capensis] were born giants who walked about haughtily and indulged in themselves every manner of theft and corruption and <u>bloodshed.</u>*

Jubilees 5:1- *- the children of men began to multiply on the face of the earth and daughters were born unto them, that the <u>Anak saw them - and they took themselves wives of all whom they chose, and they bare unto them sons and they were giants.</u> And <u>lawlessness increased on the earth</u> and all flesh corrupted its way.*

God hated the animals that the people of that time "tried" to make and the continuous wars. He called the modified animals "Unclean Abominations" and decided to finally destroy most of them, including may different variants of humans.

Genesis 6:1-7*-The Children of men began to multiply on the face of the earth, - the sons of God [Homo Capensis] took them wives of all which they chose. ¨There were giants [Homo-Gigantus] in the earth in those days; and also after that, when the sons of God [Homo Capensis] came in unto the daughters of [the pure Cro-Magnon] men, and they bare children to them, the same became the giants of old. ¨And the* LORD *said, I will destroy man, beast, and fowl.*

At the end of the Pleistocene, the Earth rotational Axis shifted, rain poured down and as the Polar Ice Caps melted and reformed, and massive unbelievable tidal waves covered the land. After the earth repositioned and the "new Earth" appeared, the following can be noted.

Australian became an isolated Island that the liquid water level rose by over 400 hundred feet according to 12 different water height studies.

Huge herds of Mammoths were quick frozen as Siberia now was in the Arctic.

A large number of the Homo-Gigantus people called Homo Capensis, Annunaki, Akamim, Archaics, Amenti, or Olympians also survived.

We find that Cro-Magnon variants carrying Y-DNA Haplotype mutations of A, B, C, D, E, F, G, H, K, L, and T and mt-DNA Haplotype mutations of L, M, N, C, D, E, G, Q, and Z survived and were found at different locations when the water subsided.

The first races of the Cro-Magnon based humans had been set before the Pleistocene Extinction. When the

Earth shifted and the boats landed, these races spread out as described in the following Haplotype map.

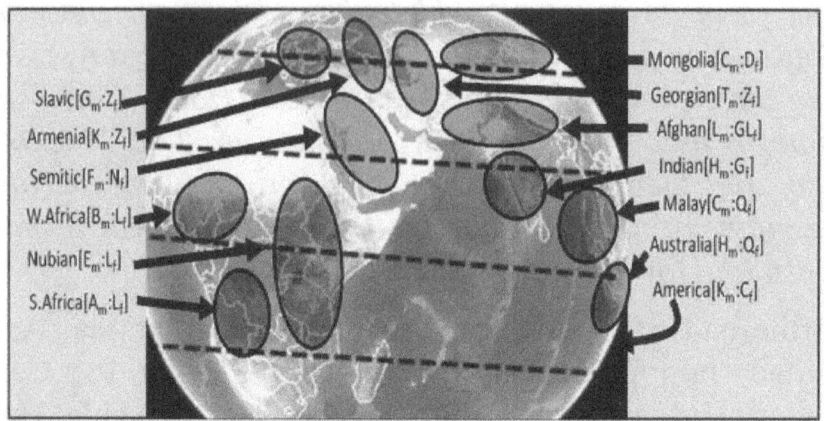

Africans- The people with the Y-DNA mutations A_m, B_m, D_m and/or the mt-DNA mutations $L1_f$, $L2_f$, and $L3_f$ would have landed in Africa once the flooding halted. This group had not integrated with many of the others during the Pleistocene Age.

Jews- The $[F_m:N_f]$ survivors were the pure blood Cro Magnon Jews. We are told one of Noah's sons [Ham] had many half breed children that filled the land of Canaan with Anakim [descendants of the Homo Capensis].

Australians- The $[H_m:Q_f]$ survivors found themselves in Australia. One of this Haplotype ancestors had been of the group known as Denisovan

Indians- The $[H_m:G_f]$ survivors found themselves in India.

Middle Easterners- The $[L_m:G_f]$ survivors found themselves in the Near East.

Even with the new information, text books continued to show that all descended form Homo-Erectus and came out of Africa. This idea of everyone evolving from

Africans got so twisted, in Egypt, Monogenism "Scientists" began changing the names on statues to show Egyptians were black men. To sort of give you a feeling for how hard people try to place the Egyptians with Nubia, the following statues are both reported to be the Theban king during the 11th dynasty, Amenemhat III. The first is easily determined to be Caucasian while the second one is certainly Nubian. By the way, these cannot be the same king. Nubian Kings took control of Egypt briefly after the time of King Solomon who provided Nubian's protection for gold. With this protection, the Nubians seized control of Egypt until the Assyrians finally took the country a little over a thousand years ago. One can assume the king to the right is one of the much later Nubian kings of the 22nd through 25th dynasties. I know you are thinking people in Egypt hate noses and smash them, but ignore that.

By scientists trying to show the Out-of-Africa phenomenon, almost all the Rulers from the very first, Narmer, until King Tutankhamun were determined to be black [at one time or another]. The collage following is an incomplete list of those determined to be black. A number look black alright, but most don't have strong "black" features at all and other examples of these same rulers have completely different features as if someone was trying to drive home the possibility that ancient Egyptians were all black.

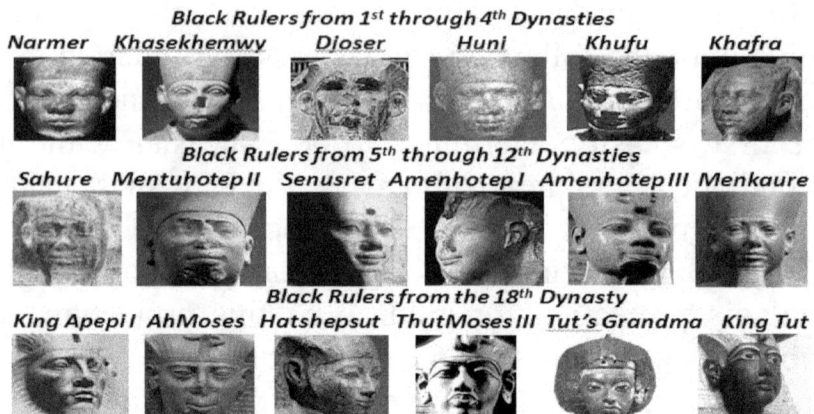

Below are the white images of Mentuhotep II, Amenemhet III, AhMoses, ThutMoses III, and King Tut---before sinister consensus scientists turned them into black people to protect the world.

Hopefully, you are beginning to realize, it is not the initial theory that causes such great harm to our understanding. Instead, it is writing, interpreting, and twisting the truth for "consensus" that is harming so many people. It is the lengths they will go to prove their statements without experimentation, theory testing, or reason and how entire societies can so easily be convinced these guys are telling them truth simply because they are scientists. Later we will describe the DNA Haplotyping of Egyptians and others to help drive this monogenetic, Out-of-Africa "lie" out of your head.

More DNA Anomaly

The full length of our DNA is made up of some 68 thousand genes that have now been identified. These genes carry the blueprint for the structure of our entire body. What is very puzzling, is the fact that Homo sapiens, as the supposed *pinnacle of civilized evolution* on this planet, should have such <u>large parts of unused DNA</u>. We seem to have the longest DNA molecule among all other species, but we use the smallest part of it in proportion to the other species. In other words, all the other creatures use much more of their DNA than humans do. Some species use as much as 98% of their DNA. We can believe more of our DNA used to be used. As we go along this will make more sense. One thing this does is, it flies directly in the face of the "principles of evolution", so the details are hidden. Think of our massive amounts of "Junk DNA" as DNA information that has been erased.

Humans should have the most complex and evolved DNA of all creatures, to have reached levels of civilization seemingly much higher than any other species on Earth over what was presented as millions of years of evolution. What is even more curious is the predicted number of genes in species. The numbers seem to increase steadily from basic organisms to the most advanced. We would expect that humans should end up

having most genes, but strangely this is not the case. Here are some examples of the predictions for total number of genes in species.
- *Fruit Fly 21 Thousand Genes*
- *Zebrafish 50 Thousand Genes*
- *Chicken 76 Thousand Genes*
- *Mouse 81 Thousand Genes*
- *Chimp 130 Thousand Genes*
- ***Human 68 Thousand Genes***

Can you see the problem here? As this supposed evolution went on, more and more genes were generated until man came along and we went somewhere between a Zebrafish and a chicken. The Chimpanzee is our closest know genetic relative and yet it has almost twice as many genes as humans. It is as if something happened to humans, but it isn't in the textbooks. And then we get to the *anomaly of the chromosomes*. Our DNA is broken up into 23 pairs of chromosomes. By comparison, all apes have 24 pairs. One would expect that *Homo erectus*, our immediate evolutionary precursor would then also have had 24 chromosome pairs so where does that put us in the chromosome development by evolution. Looks like we are somewhere between Yeast and sheep.
- Fruit fly   8 Chromosomes
- Guinea Pig 16 Chromosomes
- Yeast   32 Chromosomes
- ***Man***   *46 Chromosomes*
- Sheep   54 Chromosomes
- Crayfish   200 Chromosomes
- Butterfly 380 Chromosomes

So, let's disregard evolution for a number of reasons and look for something that isn't so silly. When we read in the Sumerian tablets that humans were "developed" as a sub-species *by* a more advanced human race [Homo-Gigantus] known as the Annunaki, it also sounds silly so we read the Bible and it says about the same thing in these 4 verses.

***Genesis 6:3-4**- There were giants on the earth in the ancient days.* [Homo-Gigantus during the Cretaceous]- *Those were the giants who were of old, men well known.*

***Genesis 1:26**- "Homo-Gigantus "elohiym" said, "Let us **make** man in our image, according to our likeness---* [Homo-Habilis-100 thousand years ago]

***Genesis 1:27**- Then God **created** man [Homo-Erectus] -- and said to them, "-- **REPLENISH** the earth-- during the 6th Age.* [Homo-Erectus humans were created 90 thousand years ago because many of the Homo-Gigantus humans died during the Cretaceous Extinction about 120 thousand years ago]

***Genesis 2:1-7**--- during the seventh age God ended His work -- and He rested – [During the 8th Age] the LORD God formed [a new] man of the dust of the ground, and man became a living being.* [Cro-Magnon appeared at the beginning of the Pleistocene 40 thousand years ago.]

The Homo-Erectus human creation was described over and over again, around the world

Mandean of Iran Story-*God first made man. When he was finished, he looked like a man, but moved on all fours, had the face of an ape, and made noises like a*

sheep. Only later did he put in a soul and teach him and make him erect.

African Erectus Mating Story- *Hairy men became human* after coming in contact with Anunnaki [Homo-Gigantus people].

Southeast Asian Erectus Mating Story -*An extremely hairy human "female" named Bota III* was cooking food. A non-hairy fisherman [one of the homo-Gigantus people] saw her and got her drunk. While she was asleep, he shaved her entire body. Only then did he find out she was a woman. She learned to wear clothes, they married, and they began a new race.

Emerald Tablets [Egypt]-*The master said-take them far across the waters until ye reach the land of the **hairy barbarians,** dwelling in caves of the desert. Follow there the plan.* [The people of Undall civilized them.]

Ngombe Tribe- *"A sky person* [One of the Homo-Gigantus people] *saw a hairy man [Homo-Erectus]. She married him and removed his hair. Then a Garden was made for man to live in."*

The Koran *[1500-year-old Muslim text]-* "*Is man not aware that We created him from a little germ? We first created man from an essence of clay; then placed him a living germ in a secure enclosure. The germ we made a clot of blood, and the clot a lump of flesh. This we fashioned into bones, then clothed the bones with flesh…"*

Laws of Manu *[3500-year-old Hindu text]- from minute body particles of these seven very powerful Purushas [Homo-Gigantus people] springs this world of men, each* *succeeding element acquires the quality of the preceding one.*

What these texts are saying is that a number of human "seeds" were created by the Creator God and the Homo-Gigantus people modified them to become more useful to them. When we look at the difference between Ape and human chromosomes we find the 2nd chromosome of a human is split into 2 pieces in an ape. Some suggest the welding together of the 2nd chromosome made us human. Many of the manipulations of DNA proved to be wrong. The image below shows the comparative similarity and difference of man and chimpanzee. Notice the 2nd chromosome.

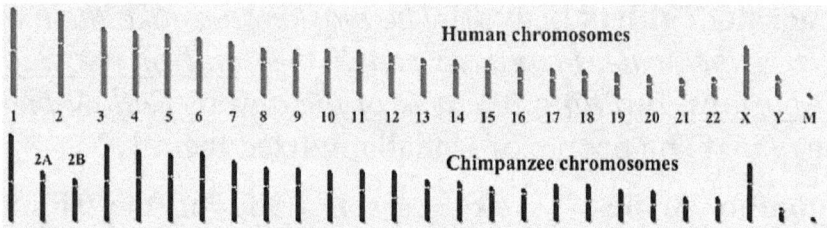

Book of Secrets *[2500-year-old Jewish text]- Those who hold fast to the wonderful mysteries of life. -With this I beseech your attention. All of the secrets of <u>manipulating life</u> were known but they [the ancient humans] <u>did not know the secret of the way things</u> are nor did they understand the things of old. They <u>did not know what would come upon them</u>---, Belial <u>who modified creation, a thing that ought never to be done again. You have not become wise in understanding my secrets of life and the earth-</u>.*
The Zoroastrian "Zand-Aksih" - <u>Satan miscreated creatures and they became useless. God saw the defiled and bad creatures, they did not delight Him.</u> Satan's

downfall was the <u>unrighteous creation of the creatures</u> and ignorance.

Zadspram- *Time made the creatures of God moving, distinct from the motion of Satan's Creatures. --After the noxious creatures died, and the poison there from was mixed upon the Earth.*

Below [left to right] are examples of Australopithicus, Erectus, Naledi, and developments which are followed by depictions in artwork from Sumeria, Greece, and Egypt of Homo-Gigantus developing various humans.

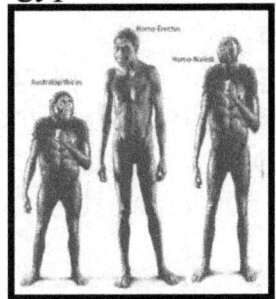

Haplogroup scientists determined the first human ancestor happened around 100 to 150 thousand years ago. This was determined by simply counting the number of replications of various genes in the Mitochondrial and Nuclear DNA. Then nothing happened for another 30 thousand years. Another major mutation was recorded after another 30 thousand years and a 4th about 30 to 40 thousand years ago. We know the first as Homo-Ergaster, the second possibly was antecessor, the 3rd one as Neanderthal, and the last one as Cro-Magnon, but something very strange happened.

Race Mutation Anomaly

What if you found out African DNA changed and had no effect on non-Africans?

Certainly, there is a strange correlation between DNA mutation and variations in race some of the traits are carried by the mt-DNA and others are supported by the nuclear or Y-Chromosome DNA. A general accounting of the basic race attributor DNA mutations is shown below. You will certainly find slight variations to this list depending on the group that makes up the description, but we are getting better and better at knowing when the mutations occur [except for the Mitochondrial mutations] and the basic location of each mutation [by using density of mutation in a group].

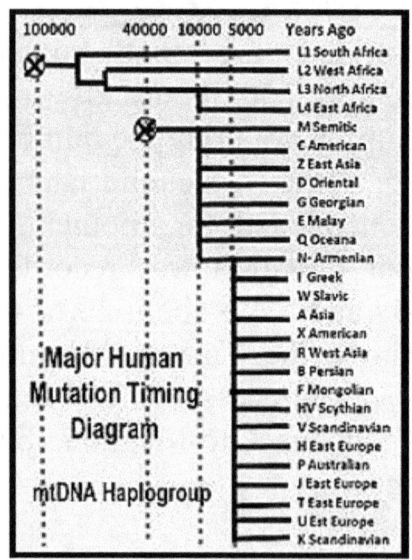
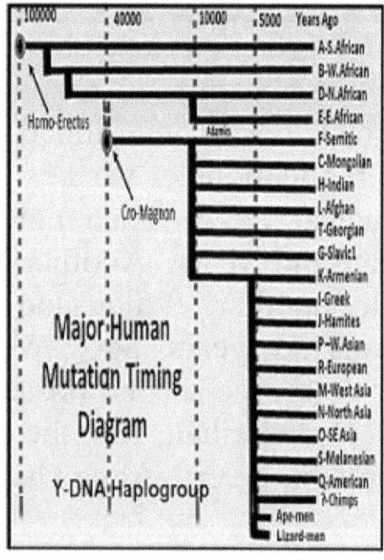

Like it or not, there is something strange about how the Neanderthal type humans changed so very much when DNA mutation studies show that there were very few major mutations during most of the Tertiary and the entire Pleistocene. Also, the mysterious Cro Magnon just popped into reality and is called, *"the most perfect human ever"*, by researchers. Others, show that the Denisovan man DNA had alien DNA and its highest match with modern man is in Australia. Every time the Judeo-Christian documentation explains something, they huff up their bottom lips and state there was no worldwide flood at the end of the Pleistocene because it messes up a consensus hypothesis. They get mad if someone even mentions the predecessor race of giants as they don't fit evolution. As fast as a paleontologist digs up evidence, they are back in the hole burying the remains in some back room. They really don't want to know what happened to humans; they only want you and their consensus group of scientists to think they are smart.

No matter what is done, almost 1/2 of all major mutations of modern humans happened at the end of the Pleistocene and the other ½ happened 55 hundred years ago and researchers should present reasons why that happened rather than hiding the fact and hoping no one saw it. Before we get into clearing up some of the anomalies, let's go back in time and look for petrified heads of Homo-Gigantus. We might call them Titans to make Greeks feel good.

Stone Bone Anomaly Cretaceous Age

How could someone find modern human skull bones that were fossilized if it takes hundreds of thousands of years to fossilize?

Besides the examination of australopithecine bones and the like to determine men came from apes, other elements of testing should be accomplished, one is to examine hundreds of ancient texts including the Bible that tell us about *"the Giants of old"* [Genesis 6] and similar descriptions from the Sumerian and Babylonian, Incan [Peru], Mayan [Central America], Ughu Mongulala [Brazil] and others that tell us about the group sometimes called the Titans or Homo-Gigantus. Certainly, we can't get DNA from stone, but what we can recognize is that these bones, some of which are huge were on people hundreds of thousands of years ago. Petrified bone should be solid proof of the Titan humans that lived with the dinosaurs, but somehow, textbooks forgot the thousands of pieces of evidence, hundreds of footprints located on the same substrate as dinosaur footprints walking along prehistoric beaches, shoe prints, massive bones, manufactured goods inside geodes, objects found deep inside coal mines.

In 1982 a Researcher named Ed Conrad discovered in Pennsylvania petrified teeth inside the jaw-like area of solid rock. Then he found more and more specimens that bore the contour of human bone. You can just call it a stone head, but petrified skulls tell a tale.

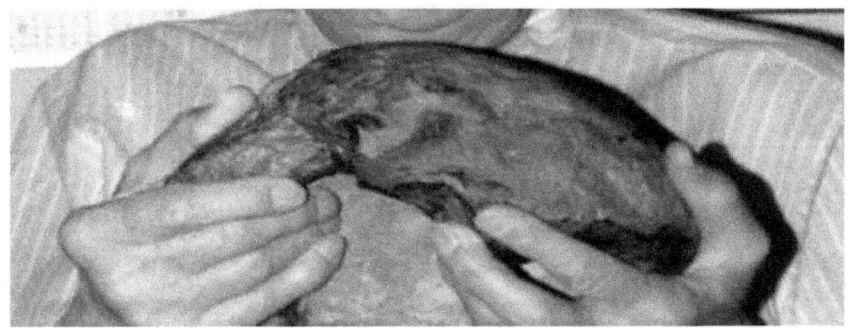

Wilton M. Krogman, the internationally acclaimed bone expert identified the first "stone bone" as a human calvarium, a portion of a skull with the eye-sockets broken off. A year later Ed Conrad discovered the large boulder in which was embedded the object that bore a distinct resemblance to a huge human cranium.

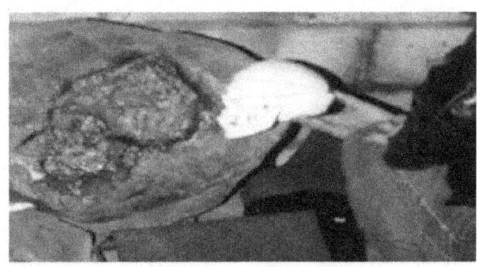 

A CAT scan had been done of this particular specimen and revealed intriguing characteristics of a human skull jaw and joints. [See above right]. I know this makes you mad that people living with dinosaurs during the Cretaceous Era were ripped out of your history books and science information that would allow you to understand how humans got here better, but there are some who will protect their "theories at all cost even if they have to bury the past as we will see in the next anomaly.

End Of The Dinosaurs Anomaly

This is actually a group of theories that are almost all messed up, but they show how Anomalies are held sacred.

While no one knows what exactly happened to kill off the dinosaurs many theories have come about to confuse and sidestep key issues in the extinction. That just isn't the right thing. Let's first look at some of the bogus theories and try to make sense of it all.

Dust Theory

This is by far the most discussed idea concerning the death of the dinosaurs. A huge meteor hit the earth and pushed dust around the world. This blocked the sun and BAM! Every BIG thing is dead. While it doesn't address the initial causes for the dinosaurs getting huge in the first place, it uses SOME physical evidence and expands it to build a somewhat absurd picture. The proof of this hypothesis lies in a layer of iridium dust particles called the K-T boundary that was deposited at the same time that the dinosaurs disappeared. This iridium stuff is fairly rare on earth, but very common in meteors so that is where the whole meteor strike idea comes from. I'll be talking about this episode later, because that part is well tested.

> *The hypothesis starts off with this huge meteor that hit the Yucatan at the end of the Cretaceous Era and spread the caustic dust everywhere. The dust supposedly choked out the light and life of our planet.*

Scientists found the layer and deposits around the world and they found the remains of a huge meteorite crater. To make the event even more scary, on the other side of the world, they found something horrifying. They found that the earth actually split open when the meteor hit and huge amounts of magma were expelled onto the surface of the earth making most of what is now called India. Hundreds of cubic miles of magma spewed up for a long time after the Meteor hit. It was a scary time. The dinosaurs essentially disappeared around this same time so there is little doubt of some of the elements of the hypothesis. The problem is that a lot of this idea makes no sense at all. First of all, most of the dinosaur bones have been found UNDER the layer of iridium known as the K-T Layer so they died before the meteor struck. I suppose one could believe that the dinosaurs had an intuition about the impending doom and were frightened to death, by the dust. It would have been nice of them to die this noble way, but, in my opinion, we still cannot completely trust this absurd theory.

Assuming that many died AFTER the meteor struck, supposedly the dinosaurs were weaker than the tiny and insignificant rodents that survived. While today's rats probably would be good survivors, these initial creatures were not stronger, more powerful, more adaptable, more intelligent, or more evolved than the dinosaurs. The new creatures that sprang to life under mysterious circumstances should have died with or instead of the dinosaurs. Besides, while the dust may have caused a slight reduction in temperature which could have brought on famine, an equally logical theory says that the dust would have most likely covered the atmosphere such that the temperatures would have increased which could have allowed for vegetation to flourish. More vegetation would have been better for the dinosaurs. Good or bad the small mammals that would generally have been eaten by the larger monsters during this time. Small mammals should not have flourished as the dinosaurs died.

To be fair, this hypothesis actually indicates that the dust layer was only part of the problem. Fires broke out everywhere, noxious fumes from the dust layer may have also killed by asphyxiation. It would have generally been a bad time that would have lasted for years as the plant-life possibly didn't flourish, but instead, would have been in trouble as well. Without an abundant plant supply, smaller animals could have had a slightly better time of it, but only slightly. In general, one can suppose that bigger is better not the other way around. Larger animals today live longer, control their environment better and usually eat the smaller creatures. In the old days, the same general theme should have applied. I know you are going to say roaches would have survived, but the delicate mammals of that time were not roaches and dust did not kill the dinosaurs. We must investigate further.

Egg Theory

> *This is another popular one. In this hypothesis, tiny mammals killed all the dinosaurs by eating their eggs.*

The proof is in the large quantities of violated nests found during the same general time as the K-T boundary, but it is pretty much known that the only reptiles that survived during this time were the small ones and that causes a problem. If one is to consider this hypothesis, he must first suppose that the other "smaller" reptile eggs were too small for the rodents to eat and that it was easier to fight off a weak, docile, and HUGE mother dinosaur than fight off a ferocious, intimidating, horrifying, tiny insignificant mother lizard. Here is where we get even stranger. Some would contend that the huge dinosaurs were too slow to fight off the mighty rats, but they would have died off much earlier as the gigantic, slow dinosaurs sort of stood around waiting for other gigantic, not quite as slow, dinosaurs to eat them.

Others would contend that the mother dinosaurs didn't have mothering instinct to protect the little ones, but the evidence shows a different story with mother dinosaurs protecting their young just like modern alligators and crocodiles. Not too many rodents eat those luscious crocodilian eggs and crocodiles are still not extinct today. Mother dinosaurs would have looked at many egg eating rodents as snacks. --- Egg eating didn't kill the dinosaurs! The mother dinosaurs must have been dead from something else that happened and the eggs were left unguarded so the mammals could feast away.

Small Hearts Theory

> *This one starts off with part of an answer, but it goes terribly wrong. This one states that dinosaur hearts could not pump blood to their bodies efficiently and that caused their demise.*

The proof lies in the knowledge that the heart cavities were not large enough to support blood flow that could support running. Without running, it is said that the hunter dinosaurs could not hunt and soon they all died. While it's hard to disagree with the premise, if dinosaurs could never run, they would NEVER have survived in the first place. Just think about a short-armed Tyrannosaurus trying to slowly reach his huge head down and get some other slow-moving food. The picture seems almost comical. Some have stated that because the Tyrannosaurus arms were so short, any fast motion would place him in jeopardy of falling over. He could not protect his fall and even if his heart were big enough, his tiny arms would have eventually killed him. The great finds of this oddball dinosaur tell us a different story. He was king during his heyday and it wasn't his heart that killed him. Let me tell you another thing. The Tyrannosaurus also ran or he would not have eaten. I know you have been told the Tyrannosaurus was like a vulture and waited for things to die before he would eat, but I tell you this is an impossible scenario. If he were too slow to kill a prey, he certainly would be too slow to get the prey after death because many other "fast" dinosaurs would have taken the food before he could get his meal. Some tell us the huge brachiosaurus was so large, even being in the water much of the time to support his weight, blood flow was an impossible situation in its miles and miles of arteries and veins. While it was impossible for the brachiosaurus to survive, it survived very well for millions of years. Its heart didn't have trouble pushing oxygen to the different parts of the huge brachiosaurus body so it could stretch its head up to the tops of the trees for the best food stuff.

Plate Tectonics Theory

This is one is gaining acceptance, but I have no idea why.

In this hypothesis, the super continent Pangea began to separate. Its separation might have somehow killed the dinosaurs as plate tectonics rammed together crustal plates enough to push dirt almost 5 miles into the sky.

While the split of Pangea has been theorized to have begun at the beginning of the Triassic era, during the height of the dinosaur rule, it is surmised that major Himalayan Mountain bumping might have occurred at the end of the Cretaceous and plate smashing would have really messed up the environment and killed the fragile dinosaurs and left those mighty rats. The whole plate movement thing is true, but the idea that plates smashed together and made mountains is absurd. Instead, the evidence seems to indicate that the 1st of the two major "world-ring" mountain ranges, called the American ridge came into existence well before a terrible catastrophe that occurred at the beginning of the Permian Era. The evidence suggests the planet Mars came very close to our planet and the gravitational attractions of the 2 planets pulled up massive sections of both planets along what had been the equatorial boundaries. The ridge runs perpendicular to the equator and along the western coastlines of both Americas and around the other side through portions of the Far East. You guessed it. 2nd of the 2 great mountain ranges, the Himalayan ridge, running generally in the direction of the present equator along the top of India and through China, came into existence when Pangea began to split because of the same thing. Mars again moved too close to our planet at the end of the Permian Era and it was far worse this time as both planets were even closer together. Mars was ripped in half and on earth the Pacific Ocean was formed as a huge chunk of our planet blasted away into space. The mountain developments and Plate tectonics were not the cause of the demise of dinosaurs nor did plate tectonics have ANYTHING to do with their development. Get that out of your head!!!

Germ Theory

Here is another that has had some followers recently.

In this hypothesis, some type of germ infected dinosaur's DNA, which resulted in their extinction.

The proof lies in the extinction of animals today as many species become extinct from viral infestation rather than environmental elements. Scientists know that major modification in DNA can be attributed to structural changes that result from the intrusion of a foreign agent into a chromosomal boundary. There are also the genetic mutations that occur from radiation intrusion as a result of a nuclear explosion. To add fire to this idea, there is a group that is investigating the almost instantaneous change in the human population some 6 thousand years ago. Apparently, around the world, humans and possibly other animals changed. They became much more primitive in capability, action, and civilization after some horrific event at that time. Whatever it was affected EVERONE by all accounts. To show how similarly this event is addressed, the culprit being investigated is some type of microbe that modified the DNA structure. I suppose it could have just as easily made man become extinct or dinosaurs. While it might have possibilities concerning extinction, it cannot explain why there are no huge animals today or why they got large in the first place, so we must continue.

Flatulence Theory

Don't laugh at this one, because some people have been afraid of high levels of flatulence for some time now thinking methane and CO_2 could somehow make our Earth too hot and the fear is spilling onto the dinosaurs.

Just like the fear of cattle, in this hypothesis dinosaur flatulence caused enough greenhouse gas in the atmosphere to disrupt environmental order which killed the dinosaurs and left only the Mammals.

The proof lies in the amount of dinosaur remains [fossil fuels] and the huge amount of cattle flatulence that have been registered and focused on concerning greenhouse gas build up. In addition to the cattle there is the dozens of greenhouse theories concerning how the planet Venus met its untimely end. No, I'm not suggesting that cows were on Venus, but if cows make a difference today, dinosaurs passing gas would have almost ignited the place. I'll look into this concept a little more, but the more reasonable thing is to not worry about cows or underarm sprays destroying our atmosphere. Possibly it smelled funny during the reign of the dinosaurs, but I think it is pretty obvious that the earth did not catch on fire. The nitrates would not have destroyed plant-life. In fact, a slight increase in the atmosphere would have been good for the plant-life. By the way, greenhouse gasses building up on Venus by themselves could not have caused the calamity that occurred on that unfortunate planet and the evidence suggests that the inferno began a mere 12 thousand years ago, but that is another story. Let's leave this hypothesis with a question. Was anyone offended by the dinosaurs?

Man Kill Theory

Physical evidence and many, many ancient texts including our Bible, confirms an extremely ancient, civilized, human existence. Later we will look at many pieces of physical and written evidence that cannot easily be ignored concerning this matter. That being said, some are trying to pin the demise of the dinosaurs on this ancient group. In this "hypothesis" the ancient humans that lived with the dinosaurs killed them off because they were too dangerous. The proof is provided in stone. Hundreds of pieces of evidence show the civilized humans lived before the extinction and it would be reasonable to assume that they would want to eliminate the dinosaurs as a threat to their lifestyle and life. Possibly, there were fears of the dinosaurs taking all the vegetation. Possibly, the dinosaur flatulence discussed earlier was becoming a problem. Possibly, the ancient man accidentally killed them with some type of DDT or similar substance. The concept of brain over brawn has always been an interesting story and concept. While there are many problems with this supposition, at least, it doesn't require ravenous and powerful rats to destroy the monsters and, unfortunately and it again doesn't explain why they got so big and why there are NO large land animals today. Besides, you would think that they would have kept a few for their zoos and entertainment.

War Theory

This one seems out there as well, but it also has some insight that we can use.

In this hypothesis, there was a huge war between inhabitants of heaven and earth and the results of the war led to the almost complete extinction of everything including dinosaurs.

Our Bible discusses such a war in the book Jerimiah, Isaiah, and Genesis. We find that the devastation was so bad, *"all the cities disappeared", "the world became like a desert", "the Earth became empty and void"*. This can be amplified. The odd part is the Jerimiah description of cities before the destruction and during the Cretaceous Period. If there were cities, there had to be people on earth before this very ancient war. All of this sounds very confusing, but we may come back to this very ancient war as we investigate further. Dinosaurs could have been killed as a result of this devastation, but again, this would not provide us with information concerning why the dinosaurs got huge and why no land animals are huge today and the pesky humans would have to have lived with the dinosaurs.

Missing a Boat Theory

This one is another common hypothesis shared by many.

In this one the only reason that dinosaurs died out at the time of the worldwide flood was because Noah didn't take them on board his boat.

The proof is in the fact that dinosaurs were here before this worldwide flood during the Pleistocene Extinction and are not here now. The problem is that they died many years before the flood not a few years before as some would believe. I'm not getting into the 5 million species of animals [assuming insects were invited] that could have been on board Noah's boat, but I will say this. If the Bible indicated that all the animals that had survived the Pleistocene Wars were saved, I believe that all that were alive at that time were saved. Some were saved in other parts of the world, but Noah did his part as well. Dinosaurs were long since destroyed before the time of Noah and there were not 5 million species of animals. That also is not a subject of this book.

Water Canopy Theory

I know many have heard about the giant water canopy that surrounded the world prior to the worldwide flood. Bible classes around the country discuss this miraculous component of our history. By one type of interpretation this canopy is identified in the Bible as the water source of the worldwide flood.

In this hypothesis dinosaurs died when the "Water Canopy" came down. The loss of this canopy affected the atmosphere or its water was the cause of the worldwide flood and it did in the dinosaurs.

A major problem with that is that the Bible never says that a water canopy was used to flood the earth and it would have been physically impossible to have the water up there without 2 things happening. The first is that the temperature and atmospheric pressure around the earth would have risen substantially as the water barrier would not let the heat escape. Many animals would not have been able to survive the horrible place. Man would have been in the, couldn't survive" list. Like the hypothesis initiates, dinosaurs would have been one of the casualties as well. The second thing to note is that no stars could have been seen through a thick mass of a significant water canopy that could flood the world. The portion of the "Genesis" story supposedly attributing the water canopy clearly states that whatever the "firmament was that separates two groups of "waters" did not restrict the viewing of all the stars and the water that caused the majority of the worldwide flood flooding came up from the deep. All these factors tell us that the flood had nothing to do with any canopy and if a water saturated canopy had ever been in the sky, it was before the time of the Dinosaurs.

Lack of Oxygen Theory

If you are still with me let's quickly get into some specific evidence trapped inside something called amber. Trapped inside the hardened goo scientists have found that the characteristics of the air breathed by the dinosaurs was still identifiable. What they found was that there was a tremendous increase in the percentage of oxygen in the air and there was an apparent increase in the air pressure as well. While there can be no quantitative and exact measurement of the oxygen, we can be pretty sure that when dinosaurs were in their heyday, they breathed easier. Dinosaurs died when the level of Oxygen mysteriously was reduced at the end of the Cretaceous. After that fateful event, less oxygen meant that the dinosaurs' huge bodies were too large to allow them to bring in enough air to support their bodies and they would have died. This certainly is not the answer, but it is an important piece of the answer. This evidence will help us track down what happened to the dinosaurs. Let's say we believe that the oxygen was more concentrated in the earth's earlier days. Certainly, one could determine the things that would have affected the atmosphere in that way, couldn't they? There are several answers and we will look as those things later. Anyway, dinosaurs did not die simply because the oxygen level was reduced. They didn't like it much, I'm sure.

More Realistic Theory

Oh My!!!! Not only have we not found the answer to the riddle of the dinosaurs, all these new thoughts are getting us MORE confused and none modify DNA. Hopefully, this book will begin to introduce elements that allow us to understand how the dinosaurs could have gotten here, why they got here, how they lived when they got here, why they were the veritable kings for thousands of years and why they disappeared and reappeared 50 thousand year later. almost overnight.

While the above snippits are nowhere near a complete list of proposed reasons for the doom of the dinosaurs, they are most of the main ones. Some of these, I'm sure you have heard and even believe while others may seem ridiculous to you. A new theory or set of theories has been established to counter the limitations of the predecessor "answers". The new view that will be presented doesn't ignore evidence that doesn't fit. It doesn't shy away from religious insight. It doesn't try to trick you into believing something that cannot work. Hopefully, it will provide you with a much better picture of where dinosaurs came from and why they are not here today. If we make a table of the various dinosaur hypotheses and try to see if they explain the major components of a dinosaur's existence, we find that none of them describe fully the dinosaur. I have shaded the areas that could probably describe the various components.

Dinosaur Story Explanations	Where They Came From	Why They were Huge	Why They Died	No Huge Land Animals now
Plate Tectonics	No	No	No	No
Earth Shift	No	No	Shock	No
Flatulence	No	No	Climate Change	No
Ancient War	No	No	Huge War	No
Water Canopy	No	Good Air	No	No
Evolution	Adequate	Adequate	No	No
Genetic Manipulation	If Ancient Humans	If Ancient Humans	Ancient Humans	No
Asteroid Dust	No	No	Power Rats	No
Eaten Eggs	No	No	Bad Moms	No
Miss Noah's Boat	No	No	Adequate	Adequate
Germ Affected	No	No	Adequate	Possible
Men Killed	No	No	Adequate	Adequate
Small Heart	No	No	Adequate	Adequate
Heavy Weight	No	No	Adequate	Adequate
Low Oxygen	No	No	Adequate	Adequate

Some of these hypotheses may give us a glimpse into some possible truth, but all of the previous conjectures are grossly incomplete but several are taught as truth to our children. If something doesn't quite fit, the testable details were changed to allow a specific hypothesis to flourish which forced a more appropriate one to be discarded and be considered wrong simply because the "establishment" had already decided to adopt a different approach and labeled the evidence that didn't fit as anomalous. This same stupidity is used to look at man's development even if it means hiding some of the DNA and completely mystify the entire existence of Homo-Gigantus in the Cretaceous Era.

Homo-Gigantus and Capensis Race Anomaly

Titans actually lived for over 100 thousand years, but their existence messed up evolution so they were eliminated from History.

To understand Erectus, Neanderthalis, Cro-Magnon, and modern man we need to understand a little bit about another human race. Greeks called them Homo Gigantus or Titans, but the Bible simply called the Giants of old [Genesis 6:2]. How these first people became the group known as the Anak [Jewish], Annunaki [Sumerian], Akamim [Mongulala of Brazil], Lord of Amenti [Egyptian], Archaics [Adena of North America], Araya [Dravidian of India] will be reviewed briefly so that you can appreciate a truer history, truer religion, and truer science than what has been spouted off by teachers all over the United States and one that doesn't simply build a huge ANOMAY pile to sweep under the rug.

Homo-Gigantus to Homo-Capensis

Both Homo-Gigantus to Homo-Capensis humans are actually the same race. During the Cretaceous they were called Homo-Gigantus, Giants of old, or Titans and in the Cenozoic Era they were known as Homo-Capensis or Nephilim. After the Pleistocene Extinction, Homo Capensis gained a number of new descriptive names described above but they really had not changed in any significant way. The following chart of human races shows the Homo-Capensis staying separate as there seemed to be an issue in cross procreation. Instead, they used DNA modification to produce a wide assortment of different people. By about 1000BC all Anak and variants had been killed or died.

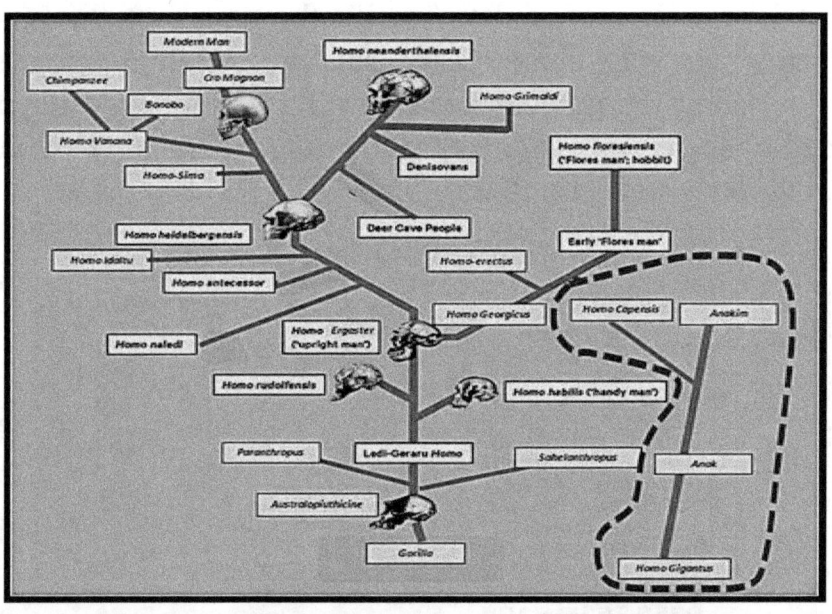

Homo-Gigantus and Homo-Capensis humans have been completely eliminated from most textbooks

Due to his size, some believe the remains of the human called Homo-Boskop was actually one of the Homo-Capensis people. I will try to keep the number of names I use for this important race down to eliminate added confusion, but do not be alarmed if Anak or Annunaki creep in from time to time.

A brief timeline of some of the more well-known modified Apes, Homo-Erectus, and the modification of that special man are shown in the following graph I showed earlier. Some of the dating in the chart may seem different than you were taught since the radiometric timing baselines have been proven to create huge error, so it was substituted for the quad-reference Ice Core, Hot-spot track, archeo-magnetic, and O_{18} density timing that is now being used to rectify the errors of the past. This in no way changes the relative timeline one can enjoy using the nuclear decay timing methods, so if you need that security blanket of timing analysis, I'm not going to try to fix on that right now.

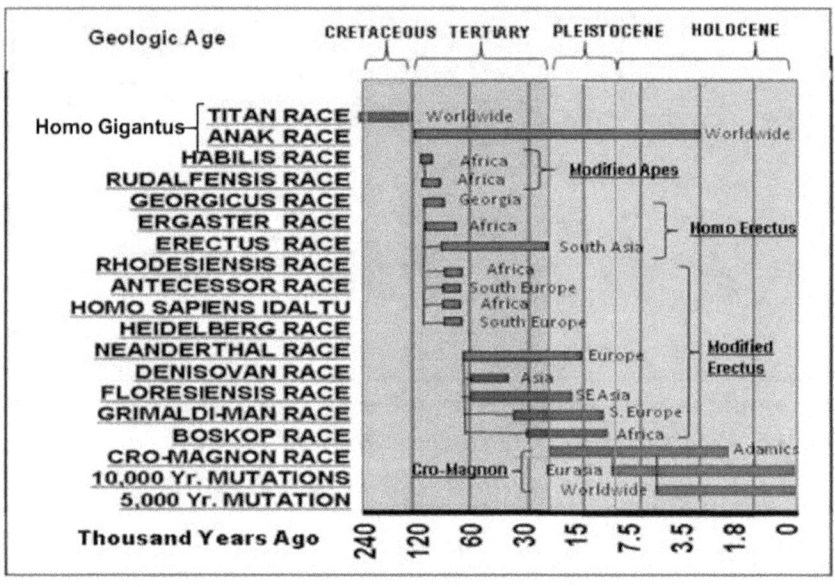

On the chart, it is the Homo-Ergaster that became the first African man and woman as described by Haplotype as the Common Ancestors. The Haplotype mutations will give us a pretty good understanding of Homo-Ergaster and Erectus and it is this same science that shows Homo-Neanderthalis [mostly found in mid and northern Europe] had very little contact with Homo-Ergaster or Cro-Magnon man who lived with the Homo Capensis people during the Pleistocene as described in the generalized chart previously shown. If you are wondering what the DNA look like, the next image is of live DNA next to dead ones.

Dead DNA- The image to the left is of the 46 DNA Strands found in a typical human cell nucleus. To the right are the same set of DNA strands after a person is "dead". What is the difference? If you give, I will tell you the answer---- "There is NO difference in the carnal existence we perceive". Dead and living DNA have the same sugars, same, interfacing bonds, same structure same chemical transfers, same biology. Don't worry I'm not getting into this area as it becomes strange very quickly.

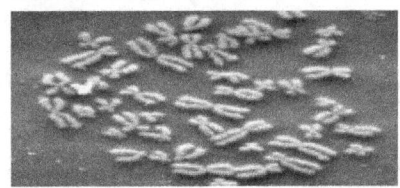 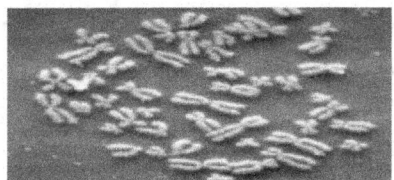

Alien DNA Anomaly

While finding Homo Capensis remains with intact DNA has not happened, what has been found are alien DNA in later Paracas and Neanderthal humans that have alien DNA [presumably from their Homo Capensis ancestors]. If we do not use the well known data, we will never understand the truth. Here are a few of the "Ignored" texts to cover up anomalies.

Mayan Creation- *"Popul Vuh"-In the beginning was total silence. God created the world and the giant gods. This first race* [of giant men] *was capable of all knowledge. They examined the four corners of the* *horizon; the four cardinal points of the Firmament; and the Round surface of the Earth. There were <u>many conflicts between the gods</u>.*

Peruvian Creation-<u>Inca</u>-*Viracocha created the world. Initially it was dark. Giants [*<u>Giant humans</u>*] were made from stone and ruled the world. The giants ignored the creator's wishes and did not worship him.*

Greek Creation-*Long after the "beginning of time", the void known as Chaos came into existence in the universe. 12 Children of Uranus [sky] & Gaea [Earth] became the first Titans. [Cronus was the bravest] A second Generation of 8 Titans included Atlas, who held the sky, and Prometheus, who helped man in opposition to Zeus]* [Some of the ancient survived the Cretaceous Era. I don't know if only 8 survived, but that was the Greek belief.] *Zeus and the other 4th generation gods came from Cronus.* [This was talking about the conversion of Titan/Gigantus to the Capensis/Olympian people who thought themselves to be like gods.]

Physical Evidence- Besides dozens of ancient texts describing these people, we are finding ancient skeletal remains of individuals 17 or more feet high. Some of the bone turn to dust as they are exposed due to such long-time periods. With hundreds of giant skeleton-finds in the United States, many are whisked away by the Smithsonian Institute and placed in storage, never to be viewed again. Hundreds of giant human footprints have been found walking along the same ancient beaches as dinosaurs and manufactured goods encased in coal, geodes, and other rocks and deposited deep in the Earth attest to the high civilization of the Homo-Gigantus.

This race of people lived for thousands of years and developed all types of capabilities we are just now rediscovering. One thing pertaining to this subject of DNA is their Genetic modifications.

Unclean Anomaly

To understand DNA and mutation or modification, we need to look at the capability of modifying DNA during the Mesozoic and Cenozoic Eras.

> *Isn't it odd that God didn't like most of the animals that he SUPPOSEDLY created?*

If you recall from Biblical text, most animals were considered "unclean or abominable", but some were considered "clean or holy". It is an absurd notion until you look into with an open mind. I know many will tell you that clean animals were the ones that could be eaten by the Adamics. Make no mistake, the Bible does indicate that clean animals could be eaten and the only thing to be concerned about unclean animals was that they should not be eaten. One problem is that <u>Noah was told to separate these two distinct creature types [clean and unclean] long before a "don't eat" definition was presented</u>. Noah, for instance was told to take 7 of each clean animal and only 2 of each abominable animal on the arc prior to the worldwide flood. The book of Deuteronomy assures us that the unclean animals were not animals that should not be eaten, but instead were animals that were true abominations so don't go thinking, I'm interpreting things for you.

Deuteronomy 14- *Thou shalt not eat any <u>abominable</u> thing. —these are unclean to you.*

An abomination is worse than just being something not to eat!!

The Zoroastrians even tell us **why** the animals were abominable.

The Zoroastrian "ZAND-AKASIH" says it this way - *Satan miscreated creatures and they became useless. God saw the defiled and bad creatures and they became abominations to Him. Satan's downfall was the unrighteous creation of the creatures and ignorance—* [with genetics].

Let's Make a List-Now that the definition is provided, let's first look at list of clean and abominable animals from the Biblical account. This is not a complete list, but has been presented to show a point.

Type	Clean	Abominable
Bird	Pigeon, Robin, Duck, Dove	Pelican, Ostrich, Eagle, Swan, Owl
Mammal	Horse, Cow, Goat, Sheep, Donkey	Camel, Rabbit, Pig, Monkey, Ape, Porpoise, Whale
Reptile	None	All including dinosaurs
Amphibian	None	All
Fish	with Fins & Scales	Squid, Eel, Catfish
Insect	Locust, beetle	All Others That Fly and all that don't

Yes, the dinosaurs were all on the ABOMINATION side. This is not a list of things good or bad. Certainly, an eagle and a swan are more majestic than a pigeon and a duck and to think of a porpoise, whale, and monkey as abominations seems strange at first. Science would think the abominations were of higher order than the clean ones so there initially is confusion.

Here are some things various ancient Jewish texts had to say about the creation of clean animals. While some fish and birds surely were "created" by God in accordance with the Genesis description, please notice that even some of the "clean" creatures **came from the earth** rather than from God's original creation. It appears, in the verses following, that genetic modification was usually done with the acknowledgement of God.

Genesis 1:24-30- *And heavenly ones said, let "the Earth" bring forth the living creature after his kind, cattle, and creeping thing, and beast of the Earth after his kind.* [This was just after God "Created Bids and Fish" that could be used. I assume the term "earth made the animals" really is "People on earth made the animals" rather than the earth itself.]

Book of Abraham 4:24-*The gods prepared* **the Earth** *to bring forth the living things. - and the gods organized* **the Earth** *to bring forth the beasts.* [This does not suggest that there were many creators during the ancient times, but instead recognized that there were very highly developed people that lived during that time. While Genesis called these guys the Great Men of Old, this calls them gods that aided in the development of different animals.]

II Esdras 6:48-*On the 6th day you ordered* **the Earth** *to produce for you cattle, wild beasts, and creeping things.*

[Certainly, God did not create them, the earth created most of the animals as was commonly known to the ancient Jewish writers.]

What Makes an Animal an Abomination?

As I stated previously and some of you are still wondering why you should trust something in the Bible, so let me go over it again so you can see the physical evidence of observations rather than a holy book of religion. what this list seems to be, is a list of animals that had been mutated by the geneticists before the flood. Before you say I'm crazy again, come up with a more plausible answer why monkeys and dolphins were abominations. Apparently, the genetic codes of these animals were so messed up that they were abominable to God according to the Bible. Other explanations don't seem to fit nearly as well, so I'm sticking with my hypothesis.

By the first chapter of Genesis one could contend that many animals were genetically altered under the watchful eye of God. My personal belief is that if God is helping you manipulate DNA the result would be perfect animals, so some of the genetics that was practiced was, more than likely, done without his oversight. If God didn't like the resulting "animals" or some major "unknown flaw" occurred because of the genetics, that could be where the abominable animals come in.

Scientists have looked around and found all sorts of strange animals and tried to piece together a crazy story that would force fit all the mistakes that were made during the genetic experiments. God would not make all the crazy mistakes noted in animals today. Eye swallowing and leg tasting animals, animals that shoot blood from their eyes, swimming penises, dinosaurs with arms that a way too short to be useful, and many other characteristics seem like mistakes because they are. Don't assume that God made the mistakes and then decided to call them abominations. He did not make them. Let's look at some of the total mistakes that had come about during ancient times. See if you don't agree that God would not have experimented this way. He had no reason to and he would not have made "Accidental" mistakes.

Evolution Anomalies

Isn't it strange that the Theory of Evolution is in total violation of the law of Entropy?

Someone came up with the Theory of Evolution. While they try to pin it on Darwin, a whole bunch of scientists are to blame for trying to support this dilemma-ridden assumption. Darwin, himself, indicated it did not work, so it's not on him. The whole concept tries to tell us that the early animals could not survive so "survival of the fittest" came along and restructured the whole animal kingdom. Besides ignoring the Law of Entropy, which would not allow the new animal kingdom to be more organized, there are the simple facts that many of the creatures that were here millions of years ago are still here with little or no differences. Clams, lizards, crabs, amoeba, sharks, alligators, many plants and other animals have had almost NO change for millions and millions of years. Certainly, survival of the fittest was not working to build these creatures, but scientists did not have to address the other alternative that many of the creatures living today and that lived in the past MUST HAVE BEEN genetically engineered. To say that genetic engineering was going on would be to say that the Biblical history and the evidence of people here long before Adam actually existed. It would explain the Clean and Unclean animals' concept and how Noah was able to get so many animals on board a small ship and it does not take away from God being the creator. He simply didn't experiment with the animals to make all the UNCLEAN ones.

Not all of the changes made by the "experimenters" were good changes either. In fact, MOST were NOT. It looks more like survival of the least fit as we compare animals that, seemingly survived.

Hippopotamus Anomaly

Short Hippo Sperm

Sorry for outburst, but I'm trying to show that the strange animal characteristics are not the result of evolution going haywire or God simply changing his mind. Even if you want to believe in evolution, there is something comical about the fact that hippopotamus have the shortest sperm of any mammal and they produce some of the largest mammals. This comedy isn't by chance. It is most likely the result of ancient biological science. The ancient humans were not creators; they were DNA recombiners working willy-nilly without understanding the consequences of their changes to God's animals. One of the changes made it easy to make wing mistakes in cats.

Cat Wing Anomaly

Wait just a minute, these wings are showing up too much. Place them in the Anomaly bucket quickly.

I know you have heard about angels having wings and possibly even laughed about it, but what about the wings being found on OTHER creatures? OK! stop laughing. Accounts of humans with wings are fairly rare, but that doesn't mean that mammals don't get wings. There have been over 130 accounts of cats having wings and there should be no limitation for the same feature being found on humanoid creatures like angels.

Above are some of the many photographs taken of winged cats through the years. From here we go to a frightened Ape.

Ape Before Ape Anomaly

Apes cannot be before the end of the Cretaceous. These dastardly quasi-scientists just made it an anomaly so evolution is not put in jeopardy.

Why does the ape shown to the right have such a horrible look of terror on its face? The reason is obvious to a casual observer, but hard to accept by the ardent evolutionist, because, the fear experienced by this mammal was because he was being eaten by an allosaurus from the Jurassic era and this type of ape was not supposed to be alive for over ¼ million years during the Triassic. The find was in upper New Mexico, but typically you don't see these things in textbooks. One of the photographs of this impossible scene is shown below and, no, it was not produced with trick photography.

These bones tell a story. The Allosaurus was ready to eat the ape. Other bones tell us even more as anomalous creature characteristics fill the "Spontaneous Evolution Theory" with doubt, uncertainty, and unbelievability.

Flying Lizard Anomaly

Isn't it strange that it has been determined that the anatomy of various creatures that lived during the Mesozoic could not have existed? Some couldn't fly; some couldn't run; some couldn't breathe.

We know that the ancient animals did do all these things, so something was different than what the other historians and scientists have been telling you. Weight was the culprit that cannot fit. The animals were too heavy to survive. If they survived, they must not have been heavy. It is as simple as that. The earth's rotation was much faster during the time of the dinosaurs so that the weight of all living creatures at that time was lower and a Pterodactyl could fly. If he could not, he would not have survived. The lizard must have been lighter, so the Earth had to be spinning faster during the Mesozoic and they forgot to tell us in the History Textbooks.

Tiny Arms Anomaly

Speaking of mistakes with arms, we have got to look at the Tyrannosaurus Rex. Previously "scientists" determined that it was the highest form of predator, but now we are not sure. His arms are too short.

They can't even reach its mouth. Although he could run, if he fell it was an almost impossible situation for him to get back up. It was possible, but the energy expended would have been tremendous and many would not have survived. The new story of the Rex came out saying that he was a clean-up animal like a vulture. Big strong, fast, tiny little arms then slow, delicate and burdened with a huge head too heavy to lift. I saw a documentary indicating that the tiny arms reduced the weight on the legs so the tiny size was necessary. Anyone looking at the huge, heavy, hulking head and saying the arms got shorter to allow the creature to stand is somehow not seeing the big picture. T-Rex was another thoughtless genetic manipulation experiment. This time the geneticists were in North America where T-Rex was supreme. By the way, the T-rex was not as delicate as some say. The earth must have been spinning fast at this time and his weight was much less than it would have been in today's earth. His lighter body made is hulking size tolerable and even useful. By the way, there could be giant people growing 17 feet tall that would be comfortable being lighter.

Strange Leg Anomaly

Let's take grasshoppers and butterflies as being strange. Their legs are used for just about everything and it doesn't seem right.

Leg Tasting-Who would create an animal **with "taste buds" on its legs**? Now it seems dumb, but when the butterfly was first created, experimentation was done on every level. *"Why not try getting the leg to taste something"*, one of the ancient geneticists might have said. Another must have said---*"Great idea!!"* and the rest is history. This inappropriate and haphazard experimentation reminds me of the haphazard DNA experimentation we [20th century scientists] have done in the 20th century that caused wings to grow in place of eyes in a fruit fly.

Leg Hearing-Then the "creators", or should I say experimenters, got to the grasshopper and decided to put its **ears on its legs**. Not only does that seem like a crazy place to put them, just think how loud it sounds to the grasshopper when he makes his leg scraping sound. I know you can probably come up with hundreds of oddities that make no sense with respect to any type of evolution process, but these are funny and appear to be jokes played on the unsuspecting animal. If you think leg ears are weird how about using an eyeball to swallow with?

Frog Eye Anomaly

Isn't it strange that a frog swallows with its eyeball?

If you are still not convinced that most animals are experimental rather than naturally selected or immediately created, let's look at the frog for a minute. I don't mean let's look at his head and say his head should be for an animal much bigger than he is. The thing that may show that it is a non-evolved, genetically experimented, thing is its eyeballs.

The frog swallows its food with its eyeballs.

Whenever it captures an animal in its mouth, it essentially squeezes the food down its throat by tightly closing its eyes and forcing the eyeball to push the food down. I have a hard time believing that God designed this anomaly and the "evolution" of eyeball swallowing doesn't make sense either. Where are the other eyeball swallowers?

Blood Squirt Anomaly

Isn't it strange that the Texas Horned Toad squirts blood out of its eye?

While we are on the subject of eyes and things inappropriate for evolution, we cannot leave out the Texas Horned Lizard. There's an evolved capability we all should have. His method for defense is to shoot his blood out of his eye. If it is sufficiently frightened it can squirt up 1/3 of its volume of blood out of pores next to its eye area according to researchers Middendorf and Sherbrooke.

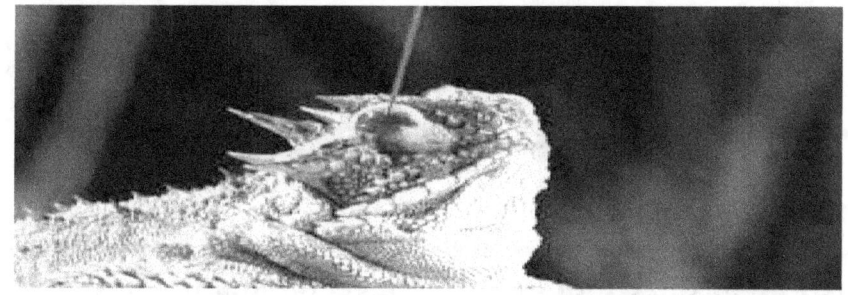

Even eyeball blood squirting isn't odd when compared to the next anomaly we call the squid.

Squid Anomalies

Isn't it strange that the 2nd longest animal in the Ocean would hide from the largest animal by staying in its mouth?

The picture above shows one of these giants. This is not a huge one, but it certainly is a respectable size, some 30 feet long. This one is in the process of dying, but it seems majestic and out of place just the same. I could go on and on about how this animal could not be generated from natural selection, but I think you have the picture. If you really want to see an inappropriately engineered creature, look at the boneless squid.

They have been around since the beginning of the Mesozoic time and they are still pretty strange. Many still have the gigantic size associated with the ancient dinosaurs. In bulk, whales are the only creature larger and whales love squid, in fact, a study determined that over 70 percent of the food that a sperm whale ate was squid. As you might expect, squid don't like this being eaten thing and larger squid have something that the smaller ones don't have. They have long talons or claws and they hold on with all their might as whales try to suck them inside their bodies.

Look at the picture of the skin of one such whale. Hundreds of sucker marks typically 4 inches in diameter and talon scratches 2 to 3 meters long typically show the vain struggle of the squids as the whales keep on eating.

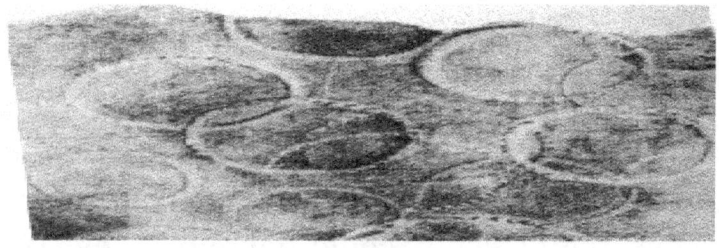

So far, they don't seem bizarre, but just wait. Giant squids don't make sense. First, they are the longest invertebrate animals in the world; although the invertebrate classification is now in jeopardy due to recent finds of jointed squid. Portions of these animals have been found which indicate sizes of over 66 feet long and 40 tons are common. Odd 18-inch diameter sucker marks found on a whale strongly indicates that animals larger than 100 feet are in existence. Even after hundreds of thousands of years of "evolution" they generally have no bones and their blood is the wrong color. It's blue instead of red. They are fast enough to stay away from whales, but so stupid that they simply try to hide in the cave-like mouth of their enemy. They hide almost all the time, but when you hide in a whale's mouth, you can expect the worst. When not hiding, they swish themselves through the water backwards which makes sure that the largest eyeballs of any creature can't see anything. They may even produce their own light to add to their strangeness. To top all this off, they are one of the most primitive and oddball creatures with respect to mating and that's where the swimming penis comes in.

Swimming Penis Anomaly

Isn't it strange that Squid penis swims away during sex?

The creatures have survived in spite of themselves. Not because of survival of the fittest, but because of forced sustainment. Let's take a look at squid sex, for instance. They can have a 4-foot long penis with barbs as shown, at least the males do, but there are two serious problems. Many male squid seem to have no sexual organs at all and no one, until recently could figure out why. To top it off, many females have no sexual opening so procreation seems an almost complete impossibility. It is now believed that squid sperm is injected directly through the skin of a female and that is where something called the hectocotylus makes its move.

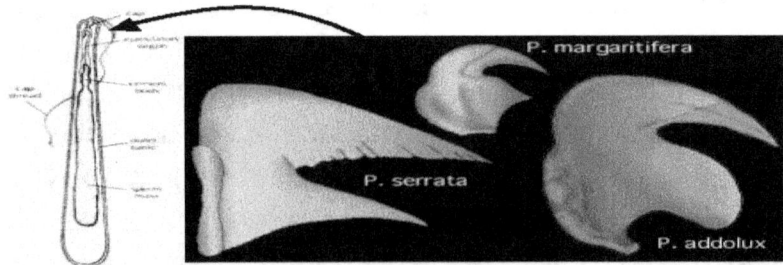

Hectocotylus Doesn't Fit-When you talk about biologic mistakes the hectocotylus should be one of the examples. These hectocotylus things were first thought to be gigantic worms of the sea. Some were 4 feet long, just like the massive organs of some giant squid.

Someone finally put one and one together and realized that some species of squid, not only inject sperm, but the entire penis shoots out of the squid so that the female can take it and have a substantial cash of sperm for later procreation. The hectocotylus [penis] actually swims to its prospective target. I don't mean the sperm swim, I mean the entire penis. If you are looking for the hectocotylus, initially, it is the third right arm of mature males. This hectocotylized arm develops in a pouch below the eye and is detachable at maturity.

One squid's loss is another one's gain. Or should it be finders keepers losers weepers?

This swimming penis thing is not a survival thing and I would have to say most animals don't have this thing. The reason is that the squid was an experiment that went wrong---NOT right. While I'm on the subject of sperm, let me discuss the odd thing about sperm noses.

Sperm Nose Anomaly

Isn't it strange that human sperm smell to find an ovum to intrude?

Few think about how basic the sense of smell really is. I know this has little to do with our discussion, but I thought you would like to know that German and US scientists have discovered a smell receptor "nose" molecule on the surface of human sperm cells, similar to those in the nose. The researchers found that sperm swim towards the smell of a substance called bourgeonal that is released by the ovum. They also found that another compound, called undecanal, blocks the effect of bourgeonal. So, if the ovum gets sprayed with undecanal, the sperm are not interested and no fertilization occurs.

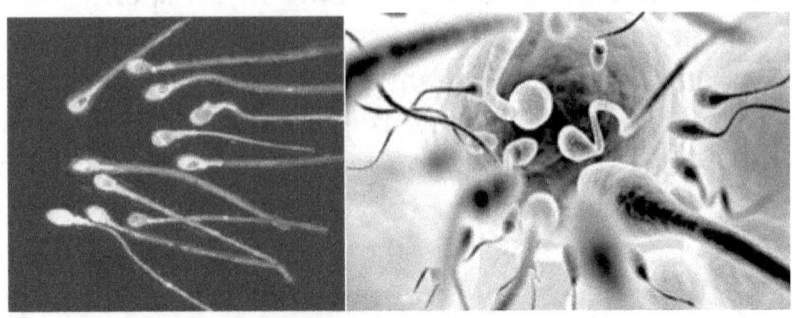

Birth control with perfume

Some are going to tell you "survival of the fittest". Only the ovum with the right scent is inseminated and no cross breeds can be had. Anyone believing that---just raise your hand right now!! OK! Some of you still believe this thing because it has been forced down you so much and you are afraid to accept these ancient people were living on earth so many years ago. Instantly, the desire to smell Bourgeonal by the sperm and the production of the perfume happened in sufficient quantities to support life. Impossible, Impossible, Impossible. You need an outside creator to begin life and you need rigorous and controlled experimentation to sustain these oddball affects. Speaking of massive changes and oddball effects, let's look at the Neanderthal human just a little bit.

Neanderthal DNA Anomaly

Neanderthal had a tenor voice and huge brain, but that messed up Evolution. This race also changed without showing DNA mutations, but that was hidden to make other theories seem reasonable.

Back in Europe we find anomalous data what we were told were dark complexion, grunting, Neanderthal. As scientists started looking at DNA from a number of sources we started to get a more definitive accounting of what Neanderthalis was like. Scientist, looking at DNA groupings now tell us Neanderthal were light skinned, freckled faced, and red-haired. They also had a high tenor voice and know about 80 words. I don't know how they determined the words thing, but for us the most important thing they found was what they called "Alien DNA". Not alien form outer space. Alien, from the Anak humans. Another anomalous detail they keep forgetting to put in the text books is Neanderthal had a larger brain than modern humans. Why change what is written so long as consensus scientist continue to support the lies.

Red-Headed Neanderthals-Ancient DNA has been used to show aspects of Homo-Neanderthalis appearance. A fragment of the gene for the melanocortin 1 receptor (MRC1) was sequenced using DNA from two Homo-Neanderthalis specimens from Spain and Italy in 2007. Neanderthals had a mutation in this receptor gene that has not been found in modern humans. The mutation changes an amino acid, making the resulting protein less efficient. Modern humans have other MCR1 variants that are also less active resulting in red hair and pale skin. The less active Homo-Neanderthalis mutation probably also resulted in <u>red hair and pale skin</u>, as in modern humans. Because this (MRC1) stuff was in Neanderthal and not in us, this gene came from someone else. Many scientists struggle with this because they will not accept the existence of the Anak no matter how much evidence shows up. By the way; some of the Capensis [Large long, and thin] skulls and associated skeletons in a massive find of over 300- skulls in the Paracas district of Peru was DNA tested. Like the Neanderthal, "Alien" genes not found in modern man were evident. I don't know if it was the same MEC1 stuff, but I think you get the picture.

Speaking- Another odd sequence called [FOXP2] was found in Neanderthal. The FOXP2 gene is involved in speech and language and this FOXP2 gene is mutated in Chimpanzee, but not in Neanderthal. Therefore, it is believed Neanderthal did not just grunt when he saw a good-looking woman. He might have said something about her eyes or shape rather than clubbing her in the head as we once believed. Also, recent physiological discoveries indicate that their voices were high pitched and nasal, not the baritone grunts we normally associate with cavemen.

Blood type- Interestingly, researchers found some Neanderthalis people had type O blood [possibly OA or OB]

Microcephalin gene –This one has been in the news lately as the news media claimed Zika virus would shrink the heads of babies making them Microcephalin. Researchers have determined that this gene comes from the Haplogroup D. Initially, it was believed Neanderthal carried this problem gene, but later studies disproved it. One can still believe Microcephalin genes came from a Homo-Erectus that was not Neanderthal.

Cro-Magnon and Neanderthal are not similar- Researchers found modern human and Neanderthal human remains from about the same time frame and found that the modern looking human had modern DNA sequencing while the Neanderthal had sequencing similar to the other Neanderthal. This showed that early anatomically modern Homo sapiens were not very different genetically from current modern humans, but were still different from Neanderthals. It was as if _Cro-Magnon was a new species_ that simply appeared and didn't change much since it "appeared".

Neanderthal and Cro-Magnon Did Not Inbreed- Many Neanderthal remains have been found in Europe and European skulls are slightly more Neanderthal-like that other parts of the world so researchers believed that would finds a link but there has been found no closeness of Europeans than any other modern human. Also, various analyses have examined the amount of Homo-Neanderthalis contribution to modern human mtDNA. The analyzers were unable to find positive evidence for interbreeding between modern and Neanderthal humans. It was as if the group that became Europeans, settled there after the Extinction of the Pleistocene Age and had nothing to do with who had been there before. This is an important point as many try to force fit colonization by location of ancient remains from before the Pleistocene event. No wonder there is so much confusion, I will discuss this more later.

Neanderthal Did Not Come From Africa Anomaly

At every turn, the humans came from Africa story gets bashed but still the text books of consensus scientist trying to hid the truth continue so their theories are not hindered.

> *Homo-Neanderthalis and modern human mtDNA is consistent with large-scale replacement and some small amount of interbreeding between modern and archaic [Erectus] populations. Interbreeding between archaic and moderns may have involved different species of archaic humans, including populations in Africa, Asia and Europe.*

It showed more similarities between non-African modern humans and Neanderthals than between African modern humans and Neanderthals but all differences are very minor. Approximately 2.5% of non-African modern human DNA is shared with Neanderthals while Africans share only 1.5%. All these things and more prove man did not come out of Africa, Survival of the Fittest is a joke; and uncontrolled Evolution does not work, but what does work is giants that lived in our very ancient past who practiced genetic modifications to produce all type of animals.

Nuclear Decay Timing Discredited

A study in 2009 indicated that Neanderthal changed only about 1/3 as much as the changes noted in Modern human DNA. This doesn't make sense in that Neanderthal supposedly lived for a longer time period. If researchers would simply get off of this nuclear decay timing, they could understand that modern humans have been around longer than Neanderthal. Using nuclear decay timing it was determined that the split that resulted in modern humans from Neanderthal occurred about 450 thousand years ago, and their heads were about to explode. One would expect with modern humans being on Earth 40 thousand years and Neanderthal living over a period in excess of 400 thousand years. Using the newer dating methods, Neanderthal only existed about 50 thousand years total so the tiny number of changes makes a lot more sense, but no one is changing the textbooks to handle the anomaly.

Neanderthal In Africa Later Anomaly

One mystery seems to be how people got where they were. Neanderthalis DNA contribution has been very scarcely found in African populations, but there is an exception when testing the non-African DNA portions of the Maasai, in East Africa. Not only is there enough similarity to show contact, DNA-ologists can tell when the contact occurred. It can be concluded that recent non-African gene flow was the source of the contribution as about an estimated 30% of the Maasai genome from about 100 generations ago. [2500 years ago- not 100 thousand years ago as some had speculated.] All that being said; a new human type showed up before the time of the Cro-Magnon rise to power. It was good old Neanderthal. It just appeared as if a scientist modified

DNA to augment the Homo-Erectus people. Again, the anomaly has been hidden away.

In May 2010, Dr. Green provided us the first Neanderthal nuclear DNA scan of about 2/3rds of the entire thing from 3 specimens which show that Neanderthals interbred with humans, and that all non-African modern humans contain 2.5% of Neanderthal genes. They also suggested that because of the high levels of Neanderthal traits in Asians that Neanderthal probably mixed with Cro-Magnon in the Middle East as <u>almost no Neanderthal characteristics are found in African peoples</u>.

Neanderthal Gene Location Anomaly

While we found most Neanderthal remains in Europe, testing modern humans describes something strange. The following map shows where people live that have more Neanderthal DNA traits that the rest of us. This includes large tracks of North, Central and South America, the Far East and Oceana. [I circled the areas of concentration so they would be easier to see.] There are almost no people with Neanderthal traits in Africa and Australia, and few in Europe.

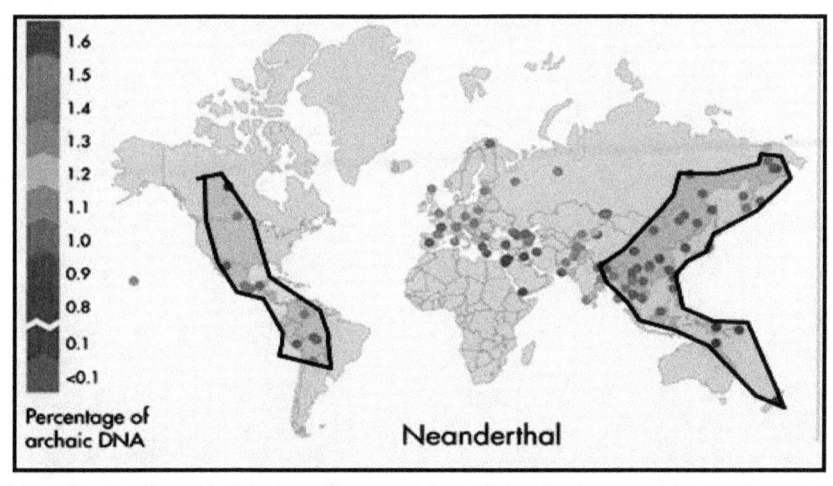

The reasons for a reduction in Neanderthal DNA traits in Europeans and the lack of African Neanderthal traits caused researchers trouble because they refused to use ancient texts to help them understand.

I spent some time on Neanderthal as he almost seemed to be the highest evolved man before Cro-Magnon came along, but then we found Denisovan. Similar to Neanderthal, this race of people had some anomalies.

Almost all die

If you are wondering about the lack of mutational change characteristics in our DNA from the Pleistocene, let me just say almost everyone died. The few survivors were not Neanderthal and the rest. They were the mixed breed of the Cain-ite "Cro-Magnon" mixed with some Homo Capensis and a few of the higher-level Pleistocene human variants. This messy group are simply called Gentiles in Judeo-Christian works, but the main thing to recognize is all survivors except for the pure Homo Capensis people were based on Cro-Magnon DNA.

Holocene Giants Anomaly 10,000 Years Ago

Called the Anakim, Akamim, Anunnaki, Amenti and Olympians, the Homo-Capensis people controlled much of the world but they messed up "Evolution" so they became anomaly.

While some of the bones described below are from the Pleistocene Era, many giants lived around the world just after the Pleistocene Extinction. While some may not have been told, hundreds of these giants have been found buried in the United States alone. While some are only 8-foot-tall others are as much as 16 feet tall and would rival the Titans and Homo Capensis of prehistoric times. While this race of humans typically is not found with viable DNA or are otherwise untested, characteristics seem to be telling with respect to common DNA. Double rows of teeth and 6th appendages are common. Red hair and fewer expansion bones in the massive dolichophallic [thin] skulls are all well documented. The Paracas humans of Peru fall in this category and some of their DNA was tested showing some "Alien" DNA as they certainly are a hybrid between the Homo Capensis and Cro-Magnon.

Turkey Anomaly

In the late 1950's during road construction in Homs southeast Turkey, many tombs of Giants were reportedly unearthed. These tombs were 4 meters long. During exhumation, the skeletal remains were examined. The human thigh bones were measured to be 47.24 inches in length. They calculated that the person who owned this Femur probably stood at fourteen to sixteen feet tall as shown to the right. A cast of this bone is shown below. Images of similar sized giants were shown in the United States, see below right.

  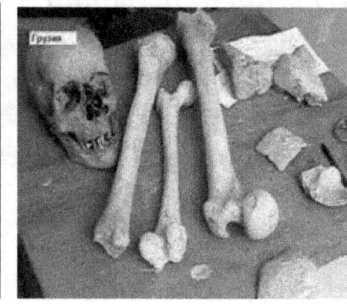

Russian Giant Anomaly

Around the world we find these Homo-Gigantus hybrid bones. The picture above right shows a normal leg bone next to one of the giants found in Russia. The middle image is from Ireland, possibly one of the Fomorian giants. He has been petrified and it was indicated he had 6 toes.

Australian Giant Anomaly

In Cosmo Newberry, In July 1970, an extensive, 4.5-kilometer-long trail of giant-six toed footprints was found about 540 Kilometers north-west of Perth. Each print measured over 16 inches long, displaying a soft pad and opposable big toe. The man's size was estimated to be over 11 feet tall. A couple of the footprints are shown next along with a skull. [It should be noted that some of the massive footprints in Australia have been attributed to the Homo-Giganticus [Ape-man] rather than a Homo-Gigantus human hybrid.

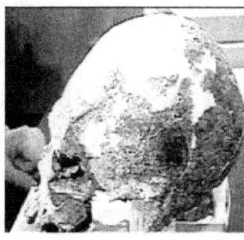 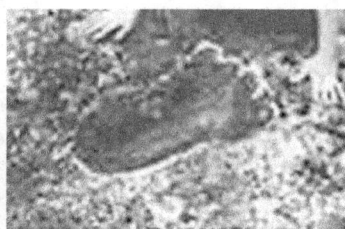 

USA Giant Anomaly

Isn't it strange that almost a hundred giant skeletons have been unearthed in the United States alone and no one is talking about the race of men that once ruled the continent?

I don't know why they make up these big words like "Dolichocephalic human", but all it means is "long headed man". This exotic species of man existed as part of a group that has been found around the world and someone came up with a huge word to identify the huge head. They were probably a hybrid group from the DNA modification by a Capensis man on a Cro-Magnon man. The drawing to the left is a depiction made of one of these giant beings. While there are multitudes of giant skeleton discoveries [the number is hundreds and increasing], here are a few to show the similarities.

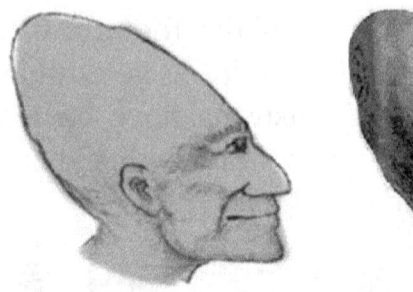 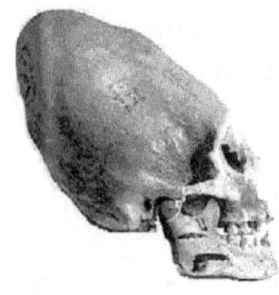

North American Long Head Giants-In Texas and New Mexico these longheaded skulls were found. Radio Carbon dating indicated that some may have lived in America as much as 37,000 years ago. Many are much younger but even these people reigned over a thousand years ago. Let's look at some of the evidence of the Homo Capensis hybrids, sometimes called Anakim. Thousands of the skeletal remains of giants that inhabited North America in the distant past have been found in just about every State. By far the Ohio Valley with its massive copper mines was the most populated. Let me give you a general view of these people. They were redhead, white people, with 6 fingers on each hand, and they had long massive skulls, some of which would fit over a "normal person's head. Some of the bones so old they crumbled at the touch and others fairly well preserved and originally many had stood over 12 feet tall. They are believed to be well over 20 thousand years old to crumble when exposed. Over 1,500 accounts from newspapers and books published in the 1800s and early 1900s describe how the United States area alone had been filled with people between 7 and 18 feet tall.

Rather than pulling in a large amount of newspaper, scientific, and historical records, here is a pretty good accounting of the giant Anak or Anakim remains found.
- Ohio - 3000 Anakim skeletons found
- Arizona- 53 Anak skeletons found-a number were twelve feet tall
- Minnesota- 36 Anakim skeletons found
- New York- 35 Anakim skeletons found
- Pennsylvania- 30 Anak skeletons found. One, found 12 feet deep was described as 18 feet long, had a 9-foot sword, an almost rusted away massive helmet, and a double row of teeth.
- Georgia- 3 Anakim skeletons found
- West Virginia- 8 Anakim skeletons found
- Illinois- 14 Anakim skeletons found
- Nebraska- 12 Anakim skeletons found
- Tennessee- 11 Anakim skeletons found
- Indiana- 9 Anakim skeletons found
- Nevada- 8 Anakim skeletons found - red-haired giants

- Alaska- 7 Anakim skeletons -massive skulls was sent back to the Smithsonian for study that were about 70% more massive than current humans
- Iowa- 7 Anakim skeletons found
- Kentucky- 3 Anak skeletons found; <u>one 12 feet long</u>.
- Missouri- 1 Anak skeleton found <u>12 feet long</u>.
- Colorado- 3 Anakim skeletons found
- Hawaii- 3 Anakim skeletons found
- Montana- 2 Anakim skeletons found
- Wyoming- 1 Anakim skeleton found
- Mississippi- 1 Anakim skeleton found
- Utah- 1 Anakim skeleton found
- Florida- 1 Anakim skeleton found, it was so ancient that once it was disturbed, much of it simple <u>became like dust</u> according to scientific record.
- California- 9 Anak skeletons found. One was <u>12 feet long</u>. A number of the remains are shown in the graphic below.

- Wisconsin- 14 Anak skeletons found [Next left shows the long skinny head. One was <u>12 feet long</u> another was estimated to <u>be 14 feet long</u>.]
- New Mexico- Anakim skeletons found [See middle below. The small jaw is a "normal" size.]

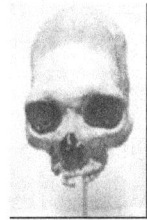 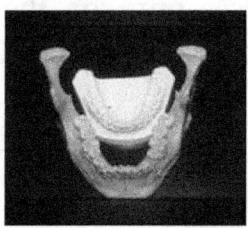 

- Arkansas- 8 Anak skeletons found up to <u>10 feet tall</u>, the last in 1976, several shown below left.
- Texas- 3 Anakim skeletons. One skull shown in graphic compared to normal skull.

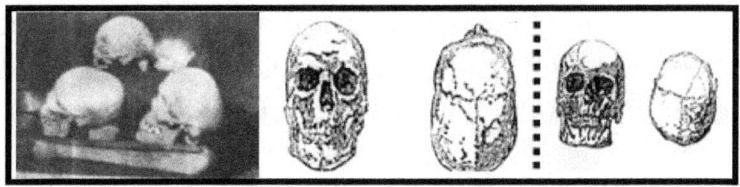

In the Ohio Valley, the ancient Adena people buried these giants under massive human filled hills as shown below.

They called these giant overlords the **Ronnongwetowanca**, but I like Anakim better. These same overlords of the Adena were also featured in the Ojibwa chronicles but were called the Archaics. It is estimated they finally overthrew these giants around 3000 years ago; however, some of the giant remains listed above were dated to be as recent as 2000 years ago.

Not only is man's brain shrinking every year, his entire body is getting smaller. Around the world we keep finding the same thing. The ruling class of people was enormous. Some reports indicate people were up to well over 20 feet tall. These guys were thought of as gods in the olden days and most seemed to have long thin heads, as you will soon see. They ruled in the Americas, the Middle East, Africa, and everywhere else in the world. The Bible indicates that normal people looked like grasshoppers next to them. Again, my beef is that school children are not even TOLD about the fairly recent rulers of the world.

South American Long Head Giants

In Peru, some time ago, a large quantity of skulls from what must have been large people was found. The people were human to be sure, but look at the skulls pictured below. Look at the long skinny head and the heavy jaw. There was no mistake that these "humans were of a different species than the other humans that lived in the area. They were Homo-Capensis.

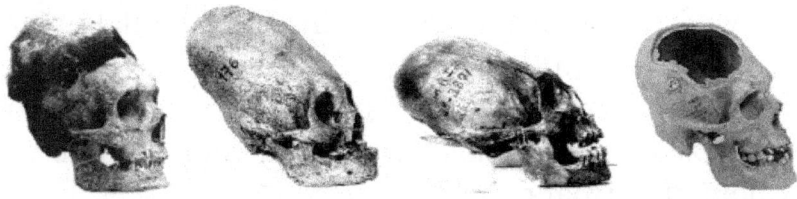

The first one must have been a character. He had a bandana around his head so tight that it indented his skull. The pronounced chin characterizes these "humans" with the Cro-Magnon humans that had miraculously appeared at the same time in the Middle East [probably a different offshoot of the Adamic humans], but these giants weren't like the other Cro-Magnon type humans at all. They were huge and their heads were huger.

Central American Giants

Isn't it strange that giants were found everywhere, but all traces of the once great society are all gone today and no one talks about them at all?

The remains of giants found in Central America have generally been the "round headed variants" or "Brachycephalic humans". Their distinction was a huge round head with the width over 85% of the front to rear characteristic skull dimension like most people living today. Like those longheaded people, researchers have found many "round headed giants". There was no mistaking these individuals from the "Normal humans around the world at that time because of their immense size. The normal Central American remains are typically about 25 to 50 percent smaller than those associated with the giant Homo-Capensis ruler remains. A pile of huge skulls in Mexico, as pictured to the left was found fairly recently. The immense size of these individuals placed them in the round category. As with most of the instances or round head finds, the giants were among normal sized human heads but segregated as if these were the rulers of normal people.

Mayan Long Heads

While we haven't found the longheaded guys, there is good reason to believe they also were in the Central American region. The Leader of the Mayans apparently had the characteristics of the long sloping head. Look at the extremely sloping skull of Pacal, the father of the Mayan nation. His ancestors possibly came from the dolichocephalic human overlords from ancient times. Certainly, his facial structure was different than the Indians of today.

Eurasian Giant Anomaly

Not only were the longheaded rulers found in the Americas, Biblical histories, and Australia, but also in Europe. Here are some specific finds from France.

*In the 17th century, in a sand quarry in southern France workers found the remains of a huge giant. Eighteen feet down, the workers found a coffin 30-feet long, 12-feet wide and - feet deep. Inside were the bones of a giant no less than **25 feet tall**. Its skull was **5-feet long** and he had <u>six toes</u> on each foot. Estimated age by the coffin depth was over 10-thousand-years old.*

In another report, we find a similar type of humanoid. "Skulls found in the Dordogne Valley, France showed a most significant trait—long heads and broad faces."

A huge skull was found in Switzerland. It was perfectly formed similar to Cro-Magnon, except it was immense and the head was long.

Egyptians

While we have no giant skeletons found in the area, we certainly can see the effects of DNA manipulation with the longhead giant rulers of the world. Look at the Egyptian statues and compare them to the Peruvian Giant skulls. See a resemblance?

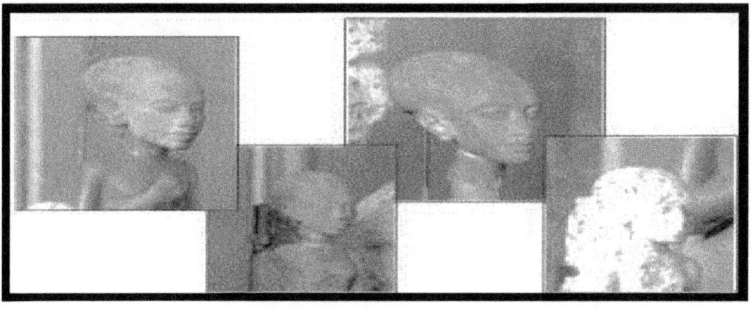

Giant Lineage Anomaly

The Chart below shows the logical progression of man. Note the giant rulers came along directly after the ancient humans, but these guys or their descendants lived and ruled over "normal sized humans for a long time. Several things should be noted. As the Genesis story and other ancient texts indicate, on three separate occasions, God made a new MAN. They can be identified as the ancient man, Homo Erectus, and Modern Man. Ancient man and modern man are pretty much identical and note that in all likelihood the Australopithecus and Homo Habilis had tails similar to other apes. The arrows show either experimentation or sexual exchange between the various groups to develop the hybrids such as Neanderthal and Homo Habilis.

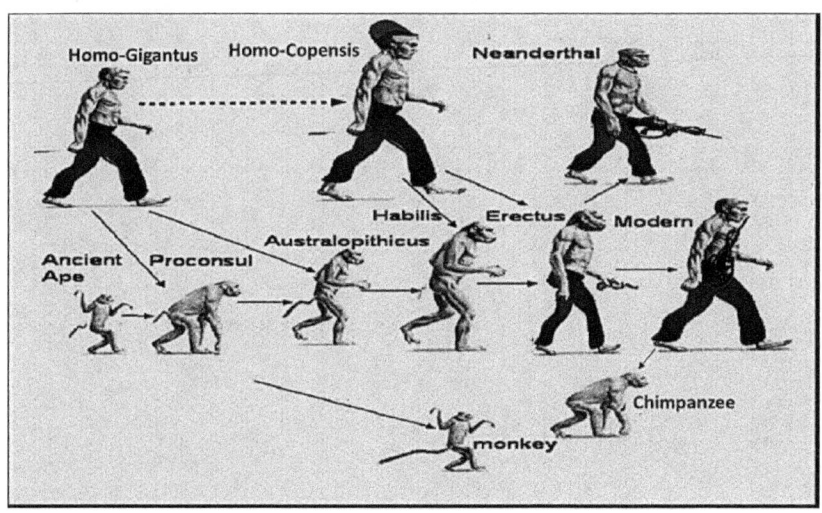

Many ancient texts besides Genesis 6 tell about how the ancient giant ancestors [Homo-Gigantus] became the Homo-Capensis human [Nephilim] and bred with "normal" humans to make other giants. One specific book "Enoch" provides substantial details concerning the religious element of all of these finds if you are interested. While I am not getting into the giants here, we should at least look at the giant animals known as dinosaurs that survived the annihilation of the Cretaceous Extinction.

Two Types of Humans Anomaly

While studying Neanderthal difference to Modern Man, it was found that there are 2 distinctly different humans. This is being kept quiet to keep people stupid.

If you look at the 2008 study again, notice there are two types of modern humans but no one is talking about the anomaly. Why do you suppose that is? How can someone show the data and then say with a straight face there is only one modern human? First let me just say Haplotype scientists try to tell us Cro-Magnon humans were not related to Homo-Erectus or Neanderthalis before the Pleistocene Extinction 10 thousand years ago but it is obvious that one of the 2 groups absolutely had mixed blood from one or more of the homo-Neanderthalis type people that lived during the Pleistocene. If we just look at the Bible description, we find that 8 survivors were "pure" while all the rest would have been Homo Capensis humans, or a mixture of Cro-Magnon, Neanderthal, Homo Capensis. The mixed breed not only made up the majority of the survivors, it also would account for the widely separating DNA mutation marks.

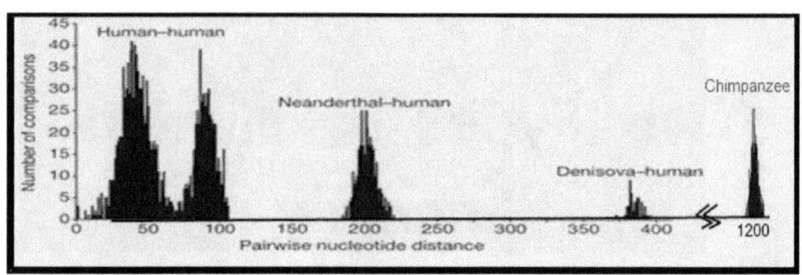

For years "undirected Evolution Theories" tried their best to meld some type of expansion disregarding DNA, but then we found out that if Homo-Erectus was the fabled 6th Age man and Cro-Magnon was the fabled 8th Age man/Adam, the story presented by Moses was unbelievably insightful and believable.

All undirected evolution ideas fall apart as they violate the "Law of Entropy" which FORCES de-evolution, but the undirected Evolution concept is pushed by the Consensus scientist converting evidence into ANOMALY.

Additional testing would make an even clearer indication that there are 2 types of humans as we continue looking at the chimpanzee. What we now know is that the African subspecies of modern man is substantially different than the Middle-Eastern Cro Magnon base human. Even after mixing for so many years, there still is a huge difference in the DNA. The differences are not as huge as those found between Neanderthal and modern man, but they are clearly different; and then there is the mysterious Denisovan.

Denisovan Anomaly

Denisovan could not have been in Asia without assistance, but he was. Let's just make something stupid up.

Let me let you in on a dirty secret. If you want to produce and sustain a viable community, you must have <u>many opportunities for procreation or you have to be produced in a lab.</u> Without controls, some don't work, sometimes location is a huge deterrent, sometimes the offspring are not viable, etc. etc. That being said, tracing back family lines to a single group from a single parent is not likely. I'm not saying there will not be similarities in how each of the potential offspring producers would mutate, but to think that all would mutate the same is--- OK I'll say it again; UNLIKELY. Somehow the Denisovan Race didn't get the information and messed up some of the Haplotype studies. Denisovan people seem to be a cross between Heidelberg and Neanderthal people; in fact, they are more closely related to the Heidelberg according to their MtDNA. Recreations are shown below.

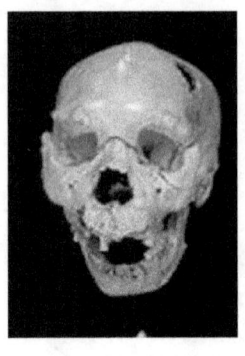  

A higher quality Denisovan genome published in 2012 revealed variants of genes in humans that are associated with **dark skin, brown hair and brown eyes** [middle image] are contained in this race showing he looked substantially different than Neanderthal [light skin, freckles, reddish hair, high tenor voice- last image]. While researchers didn't get a substantial amount of the DNA, they did get enough to provide a snapshot of the differences between Neanderthal and Denisovan. Closer than Chimpanzee, Denisovan was still much more removed from modern man. While the first Denisovan bone was found in Russia, the more interesting thing is that this seemingly close relative to Heidelberg and Neanderthal who both were European had its highest concentrations of similarity to the Australian Aborigines. I don't know what you are thinking, but I'm thinking the only way this is possible is by DNA manipulation, scientists and flying machines during the time of Denisovan. The chart below shows Denisovan is substantially closer to the chimpanzee variant than Neanderthal.

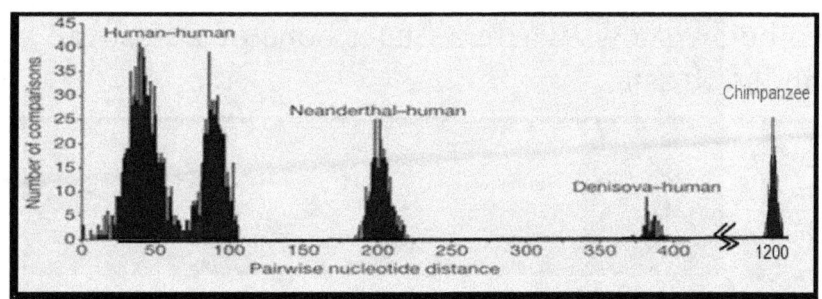

Today we find very few Denisovan specific traits, but where we find them may be telling as there are none in Africa, North America, and northern Asia. Don't worry about this map too much as the maximum similarity of Denisovan is about 0.5%. The map following shows this very peculiar distribution of Denisovan descendants.

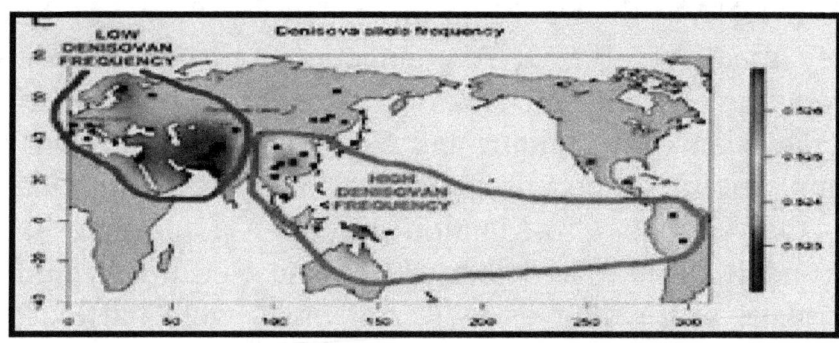

Not Out of Africa Anomaly

I know you keep hearing a new mutation of people came out of Africa 100 thousand years ago, but it simply didn't happen or the probabilities are extremely low except for the Homo Erectus people. Certainly, there were early people in that country, but there were other people in other locations as I have been presenting. One race of people was found in the Far East and SOMEHOW these guys mated with a European Heidelberg from Spain. The map below shows where a finger bone of Denisovan was found in Russia.

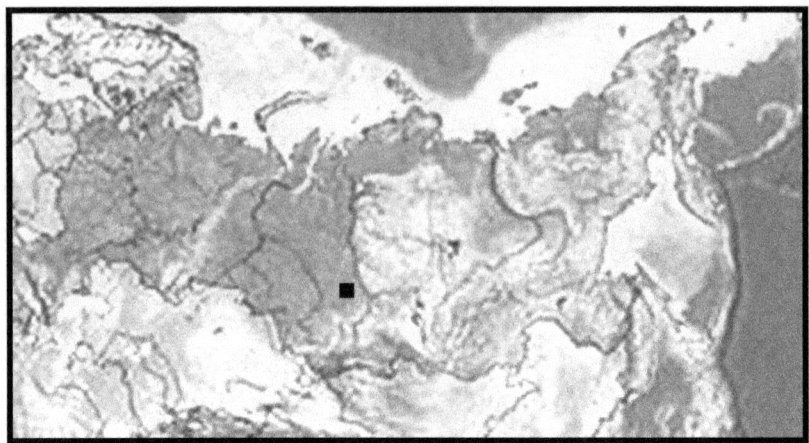

Their DNA is the troubling thing. Let me show you how they try to resolve DNA connections. In the following graphic, you will see Denisovan are genetically similar to Melanesians [Australians and New Guineans] and that is simply not good because Melanesians are directly related to the various Cro-Magnon races. Also, notice that Neanderthal and Heidelberg DNA must be linked oddly to make sense of it as well. Scientists were baffled by how this was possible. We will put sanity of these anomalies into our understanding later.

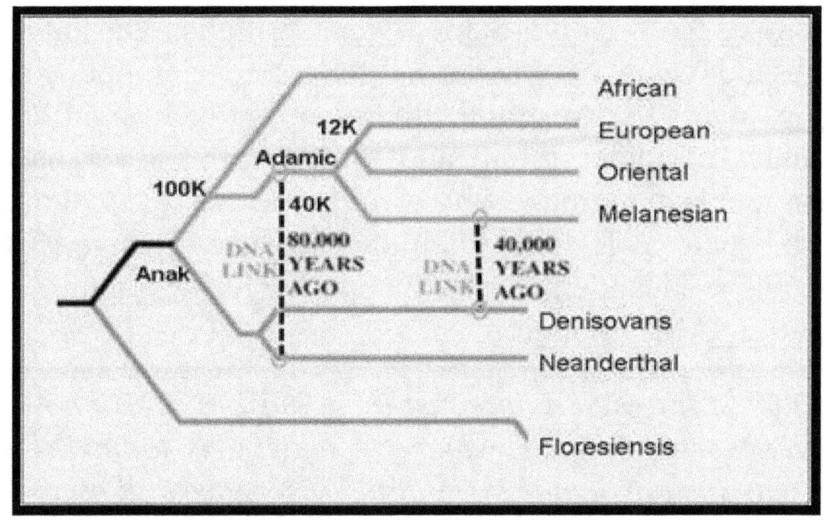

Denisovan Mutation-Denisovan were "Generally" Neanderthal that were not located with Neanderthal. Completely isolated from the REAL Neanderthal, some of their DNA changed during a mutation period differently than the European version.

Stupid Timing Anomaly

An international group of 'Consensus" scientists have completed a highly detailed analysis of DNA from what is estimated to be a 50-80,000-year-old Denisovan finger bone. Common thinking from anthropologists without insight now say about 700,000 years ago [about 90 thousand years ago by new dating], a group of humans left Africa and spread out across Europe and Central Asia. Those walking the thousands of miles to the Far East and those walking to Europe stayed the same somehow [their DNA did not mutate]. Here is where we must say there are no Neanderthals in Africa so as these people left, <u>both groups were simultaneously zapped with cosmic rays or something and both became what</u>

scientists call Homo-Sapien-Neanderthalis. During the thousand years of wandering to get to Cambodia and jump over to Australia, there were NO significant changes of either group and the Africans never had a cosmic blast to make any similar people even though there such a HUGE likelihood of the change from Erectus to Neanderthal that <u>2 completely separate groups had the SAME mutation.</u>

Oops! I laughed a little just then, but I'm over it now. This is what happens when a scientist gets so involved with his particular insight and disregards all other information around them. No one even thought to ask how the finger bone got to Siberia.

Besides this guy that must have taken a wrong turn to go North, soon they go tired of walking and had kids. The kids of both groups stayed the same. Both are Neanderthal except for the unusual DNA that somehow was introduced. These people became Neanderthals and Denisovan, and the people they left behind in Africa or the Middle East became Homo-Sapien-Sapien. So how does a Denisovan finger bone illuminate what happened to Neanderthals? <u>One thing that we now know is there are 23 known areas of the genome that modern humans do not share with either Neanderthal or Denisovan.</u>

Flying Anomaly

Another thing that is almost certain is that flying transports were common in the olden times allowing for widely spaced separation of individual, societies, and trade. Hundreds of document, physical evidence, and artwork describe them, explain how they were used and allow us to understand more about DNA if someone would simply try to explain something in a less comical way.

Cross-breeding-Scientists have now found a crossbred individual "half Neanderthal and half Modern man". When the DNA structure of Europeans is examined, it is found that they now are between 1 and 4 percent Neanderthal. When I say cross breeding, I have to also add in Denisovan.

Oldest DNA-In an underground cave in the Atapuerca Mountains in northern Spain, a pile of bones was found. In the pile was a leg bone with DNA. The bone is apparently 400,000 years old [old dating] or about 50 thousand years old [new dating] and shown below.

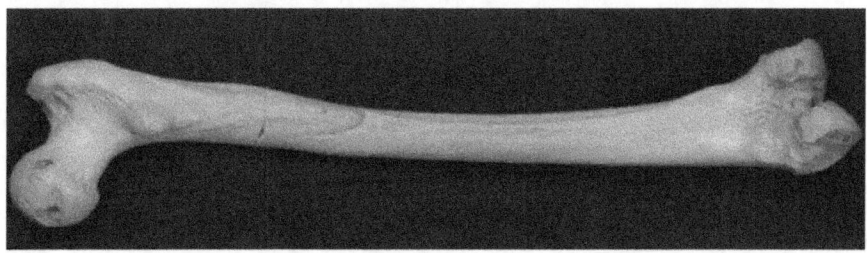

The researchers reconstructed a nearly complete genome of this fossil's mitochondria. The fossils unearthed at the site resembled Neanderthals, so researchers expected this mitochondrial DNA to be Neanderthal but to their dismay, the ancient Mother was Denisovan.

Now it seems that these Denisovan stayed with Neanderthal and went northwest to get to Spain. Then the Denisovan group was mutated without the Neanderthal getting exactly the same mutation and most of the

Denisovan left for Australia while at least the finger of one of them went north to Siberia.

Please don't get me wrong. I don't believe this story. It is what they are trying to make everyone else believe.

Homo Capensis Inbreeding Anomaly

Alien DNA in Neanderthal and Paracas is generally being ignored.

In 2010, it was concluded that the Denisovan population shared a common branch with Neanderthals or Heidelberg from the lineage leading to modern African humans. This suggested that the divergence of the Denisovan mtDNA resulted from the persistence of a lineage purged from the other branches of humanity. A detailed comparison of the Denisovan, Heidelberg, Neanderthal, and human genomes has revealed evidence of a complex web of interbreeding among the lineages.

Through this interbreeding, 17% of the Denisovan genome represents DNA from Neanderthal population, while evidence was also found of a contribution to the Denisovan from an ancient human lineage yet to be identified. I'm not saying this "alien" DNA was Homo Capensis DNA-----Maybe I was!

While some of this stuff has been discarded from our modern textbooks to protect some unfounded, half-baked, theory or claim, I am certain no one can hide the fact that our brain has been shrinking since for the past 5 thousand years.

-----Well, let's review this detail just in case I'm wrong.

Huge Brain Anomaly

Our Shrinking Brain is ignored so that Ancient Nuclear War can be ignored.

The earliest known remains of Cro-Magnon are about 20 to 40 thousand years old, while Homo-Erectus claims ages from the tertiary period and there is little doubt that Homo-Neanderthalis came from Homo-Erectus, however there are strange mutations in Homo-Neanderthalis that made him more intelligent, stronger, and more Cro-Magnon like. We can assume these mutations were from the Homo Capensis people. All that being said, coming from the Middle East, Cro-Magnon and subsequent hybrids are found all over the world with a curiously small population in Southern Africa. Cro-Magnons were robustly built and powerful. The body was muscular; the forehead was fairly straight rather than sloping as found in Neanderthals and he had almost no brow-ridge. The face was short and wide and the chin was prominent. <u>The most interesting thing about Cro-Magnon is his average brain capacity was about 1,600 cc; about 10% larger than either Homo-Neanderthalis or modern man.</u> [See Graph]

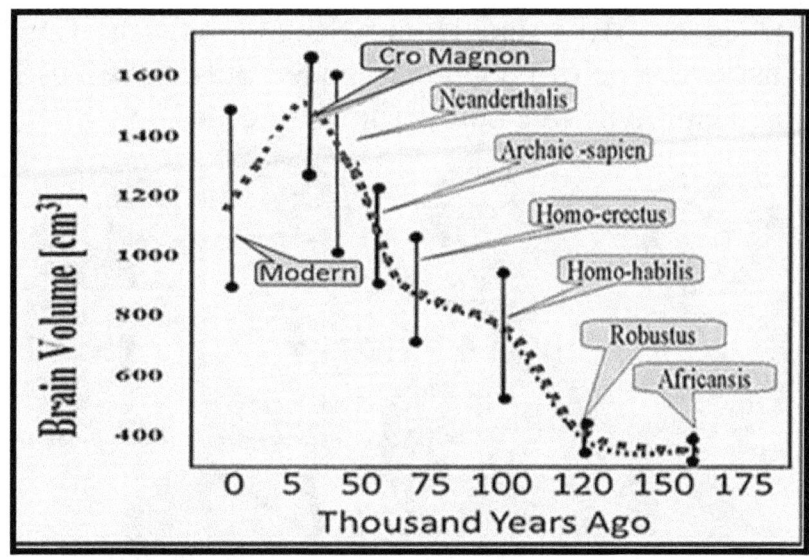

I was pulling your leg earlier. This fact is either labeled as anomaly or simply ignored so no students might ask questions. Later we will talk about what happened 5 thousand years ago to make our brain atrophy or shrink, but right now let's continue. Cro-Magnon was crafty from the start and they have been found with numerous tools along with pieces of shell and animal teeth in what appear to have been pendants or necklaces. Great artists, works in ivory, and wall paintings were popular. They also buried their dead intentionally showing knowledge of ritual and healed wounds show they protected family members.

For the next detail, I must tell you a little more about Haplotyping. Essentially it uses mutation position and type to classify and time the various elements of an individual's heritage. As I previously developed, Cro-Magnon people are identified by Y-Chromosome DNA mutation [F] and mitochondrial DNA mutation [N] "Haplotype". This is written [F:N]. Homo-Erectus is identified as [A:L] and Homo-Neanderthalis as [D:C1] the last two type of people descended out-of-Africa. The simple Haplotype tree below shows lineages.

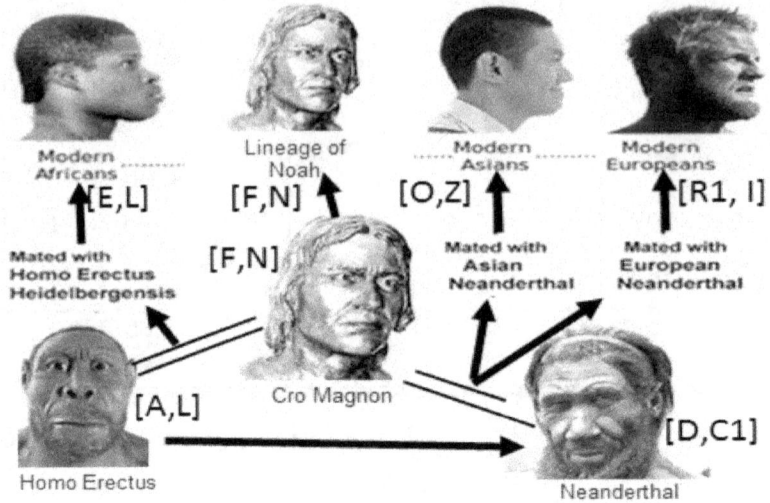

In a resent Haplotype study of almost 8 thousand individuals showed European lines had no A or B "Homo-Erectus" DNA mutations showing there was no early African link and almost no Neanderthal.

No "Out of Africa" DNA Dilemma

The Bible indicates a third human was "created" during the 8th Age. Scientists tell us Cro-Magnon humans simply appeared one day at the beginning of the Pleistocene Age well after the Homo-Erectus humans of the Tertiary period. Neither science nor the Bible can separately answer us adequately, but together, the answer becomes clear. DNA tells us Cro-Magnon did not "Come out of Africa" so from where did he come? In fact, the researchers made note of their repeated absence stating---

> *"Not one non-African participant out of more than 400 individuals in the project tested positive to any of thirteen 'African' sub-clades of Haplotype A".*

If Neanderthal did not mutate into Cro-Magnon and none of the new Neanderthalis type humans came out of Africa, how were they made. How did Denisovan-like humans get to the middle of Russia without making babies and how can Australians be more Denisovan-like with European samples being found? Unfounded accusations of racism have become common as the prevailing Afrocentric hypothesis is constantly being challenged by the growing mountain of conflicting scientific evidence, especially in the evolving field of genetics. All Cro-Magnons are modern, but all modern humans are not Cro-Magnon. In the Judeo-Christian account, there was a large number of scientists during the Pleistocene Age that the Judeo-Christian writers and others indicated, worked on Genetics. If we look at haplogroup maps we can see the anomalies.

X- Mutation Anomaly

If you are wondering why no one told you there were two types of modern humans, I can't say I blame you, but here is the really sad thing. There are many in the labs testing DNA that know about this difference and they must be afraid to describe it for some reason. Speaking of not telling you things in school, the next Haplogroup map anomaly shows the density of populations with certain mutations. I picked the X mtDNA Haplotype mutation for this example for one thing. The map shows that during extremely ancient times, North Americans came from a group in the Middle East somewhere near Israel and they never left. We can believe they were gentiles with Neanderthal, Homo Capensis, and Cro-Magnon blood.

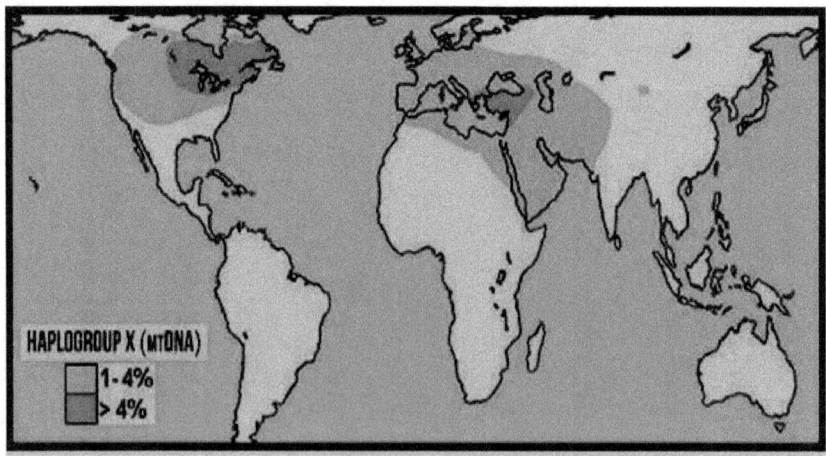

As the Americans could not have gone through Asia, they must have gone over the Atlantic Ocean. Oop! You weren't told that in school nor is it written in the text books being used, but that does not change the Haplotype maps being developed and refined every year and always showing the same thing.

Land Bridge Anomaly

Now for the important part that I have stated a couple of times previously; they did not come across that stupid land-bridge from Asia. Besides this one example of teachers not telling students the truth or the information that could allow students to think, I can show others but they all say the same thing the X mutated Cro-Magnon came to America 10 to 20 thousand years ago in a direct flight from the Middle East. They did not walk and have babies along the way as they passed Asia into Alaska or anything like that. They must have flown here or they had great ships during the Pleistocene. To help with all of this, let's look at a little more DNA Haplotype tracking. The tracking seems to have issues in reasonableness, but we will make sense of it better in a later chapter.

Homo-Erectus During the Holocene

What we will see as we dig deeper is that humans had almost no mutational differences right up until the end of the Pleistocene Age. This seems strange given the massive number of different races of people that have been recorded. Right now, we need to check out the track of mutation from Homo-Erectus. [A_m:L_f] DNA. What we will see is that this first human was in Africa 100 thousand years ago. He moved around a little for 5 to 10 thousand years and then a major mutation called Haplotype [B_m:L_f] occurred, but generally this person also stayed in Africa. At about the same time [C_m:L_f].

People with traces of Y-DNA Haplotype "A" or "B" mutations or the mtDNA "L" mutation seems to indicate a Homo-Erectus human was way back in their lineage, but do not misunderstand something important.

No Homo-Erectus survived the Pleistocene Extinction. This means no Ergaster, no Heidelberg, no Denisovan and no Neanderthal.

They all died during or before the Pleistocene Extinction. Only three groups survived, Pure Cro-Magnon, the Homo Capensis People and half-breed mixtures of "homo-erectus variants", Homo Capensis, and Cro Magnon together to make Gentiles. We know this because modern human DNA structures are all similar. If they had been based on Neanderthal, the alignments would be different. As I mentioned before there are actually 2 different types of humans walking around. One is more Cro-Magnon while the other is more Homo-Erectus or Neanderthal. Today, because of mixed breeding, there are no pure Cro Magnon and the last of the Homo Capensis people died about 3 thousand years ago. Today, we are all "gentile mixtures" with Neanderthal, Cro-Magnon, and Homo Capensis blood. Even the Jews all married outside their lineage to become gentile. The entire book of Leviticus was made specifically to keep the original Jewish people isolated from everyone else [Gentiles and Homo Capensis]. It was impossible from the start. We are told that Ham son Canaan was the first to "stray" and Noah cursed him for it. We even know his Homo Capensis wife who was named Semiramis. She and Canaan had a boy named Nimrod who would be a prominent general during the Bharata War, but that is another story. If we get back to the Jews, they had 612 rules that mostly addressed separation from the rest of the world. They did not live by them and soon all of the people of the world were "mixed gentiles".

Let me get religious just for a second. The problem with being a mixed breed is that our "spirit" part of our Triune self [Body, Souls, and Spirit] is flawed and the only way to live in heaven after death is to get it fixed. This was done by an incarnation of our creator God and that's all I'm going to say about that. What we need to recognize with all this mixing of races is that Haplogroup technology tells us exactly where and when we came. This is uncomfortable for historians so they simply leave out the details. Here are a few of the many anomalous maps.

Q Haplotype Anomaly

The first one is the mutation group called "Q". It is almost exclusive to South America and Western America as shown in the map below. What the map actually shows is that after many centuries, the people living in the Americas crossed the Bering Straits into Asia; not the other way around. They truly appeared one day in South America as either the PreInca of Peru or the Ughu Mongulala people of Brazil or they were both the same people. Please be thinking to yourself that flying transport was common during the early Holocene.

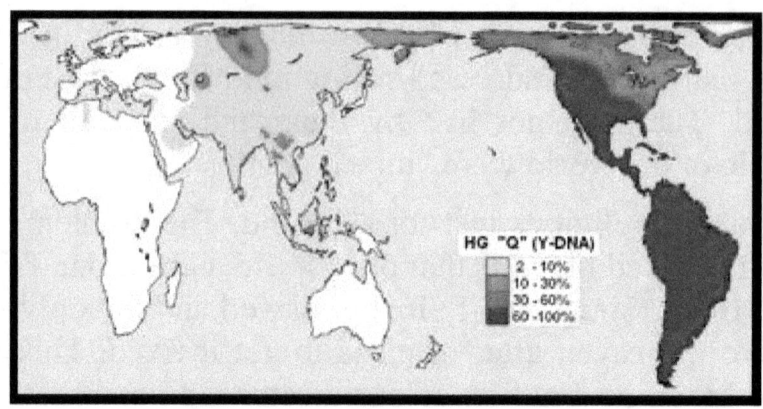

N Haplotype Anomaly

The next Haplogroup map shows distribution of the mutation N which is almost exclusively found in the people of Sabiria. It is as if life simply sprang into existance from nowhere unless early Holocene people had much more suffisticated machines for travel than provided in our textbooks.

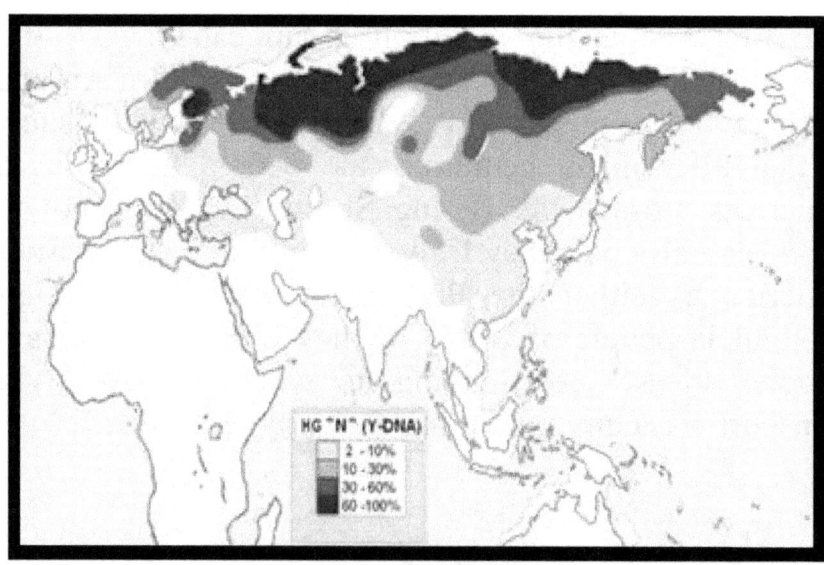

T Haplotype Anomaly

Native Americans are conventionally held to fit into a handful of haplogroups. Haplogroup T is not among the haplogroups most geneticists recognize as Native American but they have just as much of a problem in timing as the X and N Haplotypes. The map to the left shows how the densest populations of T haplotype is in Egypt along the Nile and in Arabia up to Turkey.

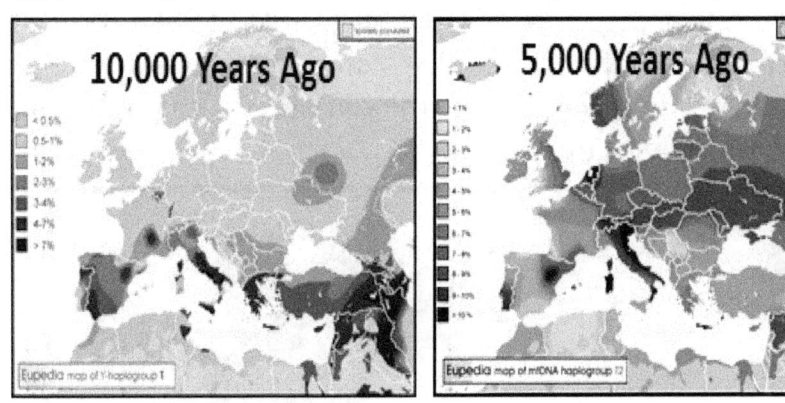

While T is mostly associated with Egyptians and Arabs "T is the leading haplogroup with over 23% assertion, which is similar to Jews, Egyptians, and Arabs with almost the same percentages. The second map above shows a later mutation haplotype T2 which occurred around 5000 years ago so we know the Cherokee nation came to America around the time of the Pleistocene extinction. Some would say they flew over, but that is just me.

J1 Haplotype Anomaly

Let's look at one more before I get into details of haplotyping and how our history textbooks authors would rather die than admit their entire works are made up fiction so they eliminate all truth as anomaly. This haplotype along with the T [Georgian] haplotype we look at previously make up a majority of Egyptian heritage, as shown below, on the east side of the NILE is it almost totally T and J1 while on the other side of the River, the characteristic African bloodlines abound. J1 haplotypes began to be seen around 5 thousand years ago so we see there was initially T haplotypes in Egypt and J1 influx from the Middle East during the time of the Jewish occupation of the land of Goshen (1800-1550BC) on the Eastern side of the Nile. As stated, before the ancient Khemetian Egyptians and the Theban Egyptians had Semitic heritage not African.

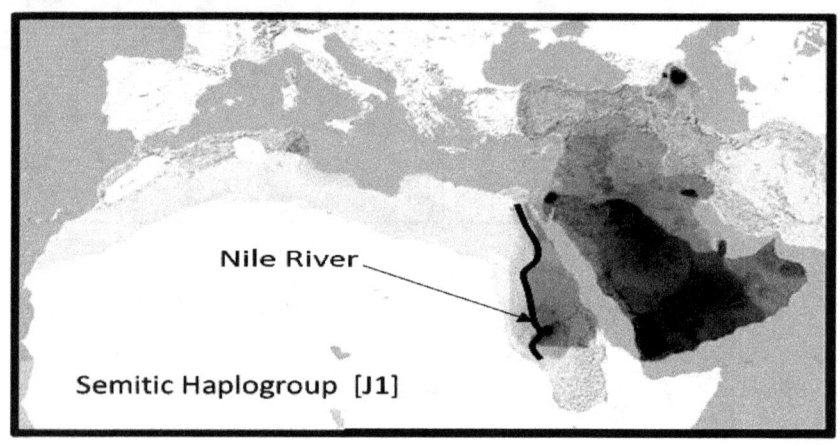

Haplogrouping by Y-DNA

For this we will again look at the Male chromosome DNA mutation Haplotypes that mysteriously occurred near the end of the War.

- L-Dravidian India- Dark Indians spilt from "H"
- M- Asian Mongols split from the Haplotype "K"
- P-Proto-Western Asia split from Haplotype "K"
- Q- Northern American Indians split from "F"
- R- Balkan people split from "P"
- R1- Aryan reddish skin colored came from "R"
- R2- Mid European white skin came from "R"
- S-Melanesian Orientals came from "O"
- T-Georgian Middle Easterners came from "G"
- Vanara- [only lived from 3100BC until about 2000BC]
- Bonobo and Chimpanzee

Patriarchal DNA Mutation and Expansion

Let me try to put perspective in this entire Y-Chromosome DNA mutation thing by showing you where the initial concentrations of the various mutations were. The Adamic "F" people were here before the flood, but others were also here. Note that Cain left the main group of Adamics and went east and the "G" chromosome mutation seems to be this group. The A, B, C, E mutations are mutations of the Homo-Erectus variant group while F and G describe the placement of Cro-Magnon.

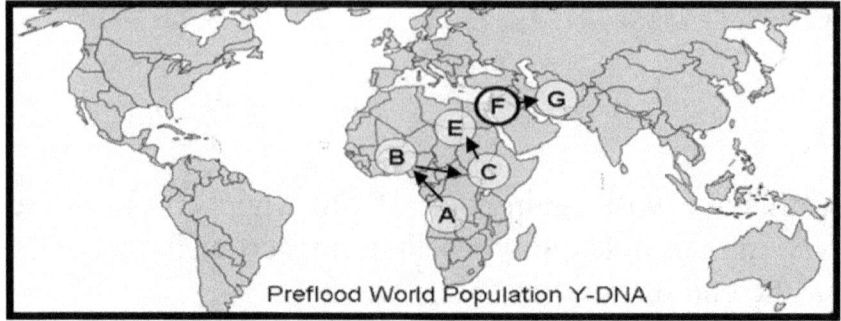

Survivors of the Pleistocene Extinction and flood had mutations identified previously and as shown next. The A, B, C, and E mutations stayed in Africa, but most of the remainder scattered throughout the world.

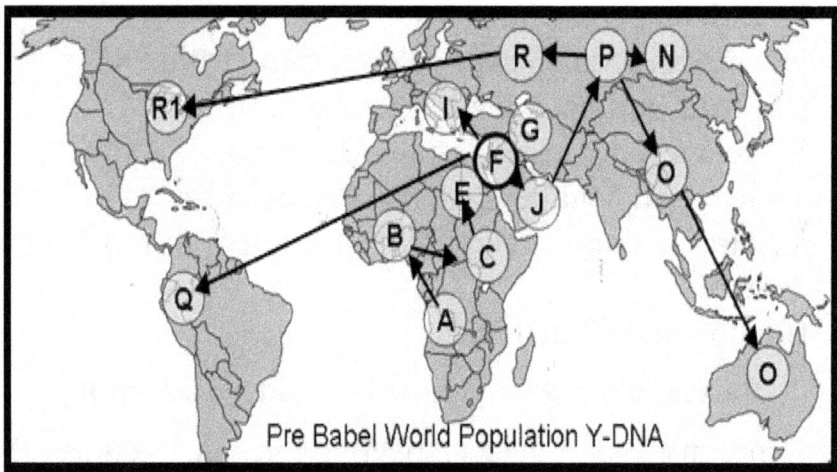

After the Babel War or Bharata War, we see many new mutations as expected.

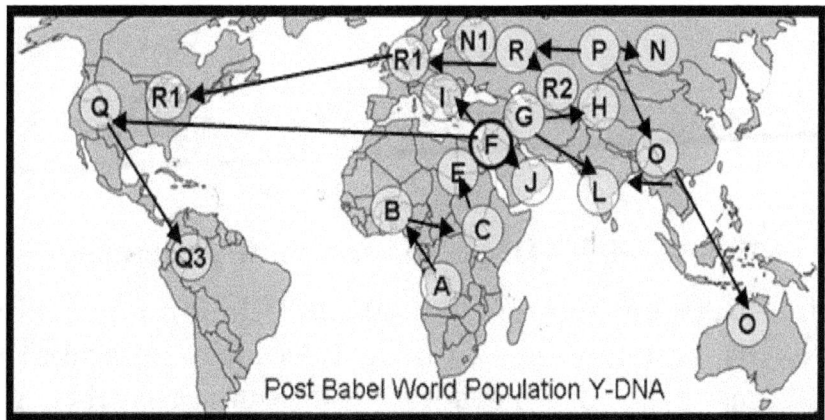
Post Babel World Population Y-DNA

As described here 2 major conflicts caused massive mutation to all humans. The first was near the end of the Pleistocene; the second was a massive war that preachers and historians seem to ignore no matter how much information would reduce uncertainties of why we cannot use our brains and idiot savants, somehow can do miraculous things as if, at one time all people could use their brains for miraculous things.

Haplogrouping by mtDNA

Mitochondrial DNA Mutation and Expansion

I didn't discuss too much about the Alien mitochondria DNA in the body, but we believe Mitochondria used to be some type of "intruder" into our cells that got comfortable and simply took up residence now it is found in almost all cells. Because of this strangeness, Mitochondrial DNA is completely different than the Y-Chromosome DNA I have been tracking. I thought showing the same maps might give just a good a picture. What we find is not surprising. The Adamic people "M" and "N" were here before the flood, but they certainly were not the only ones. The L mutation in Africa [Homo-Erectus] was here during the time Cro-Magnon M and N mutations came along.

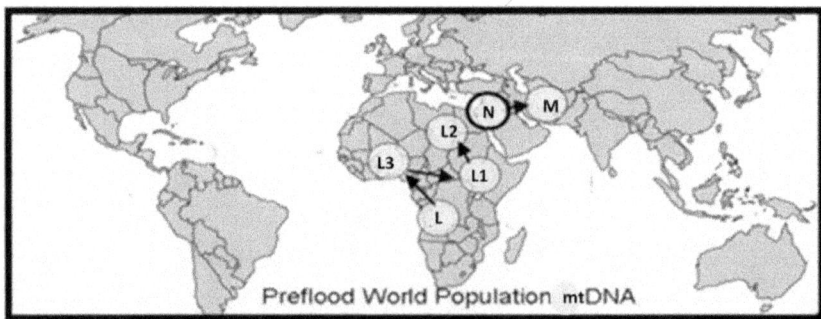

At the end of the Pleistocene Extinction, survivors were found around the world and massive mutations occurred as shown in the Americas, Europe, Asia and the Orient.

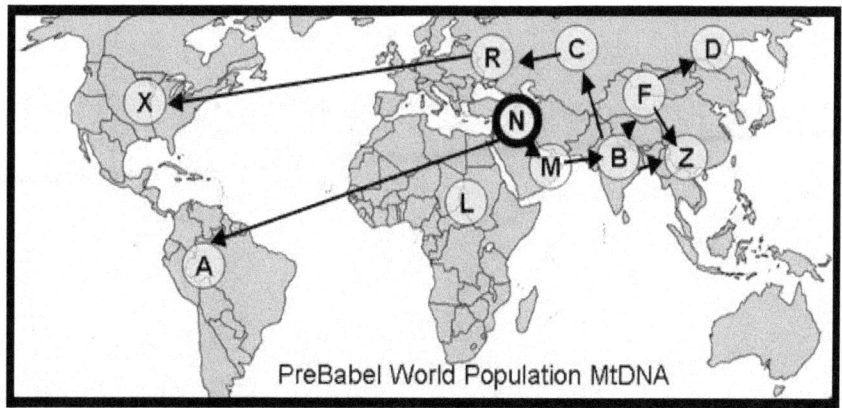

After the Babel War, we find more mutation and more separation showed substantial worldwide expansion as shown next.

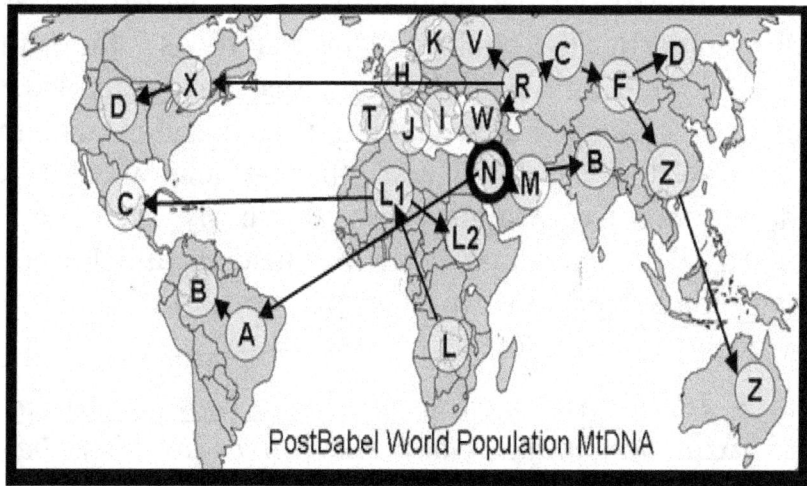

Most of the main mutations seem to have occurred during this horrible war that left us unable to use much of our brains, but there is another telling marker that shows about this mutation. People, around the world began living shorter lives, but first we need to see who survived the flood.

Pleistocene Extinction and Survivors [7000BC]

The time was 10 thousand years ago. During the extinction, the Earth shifted on its axis and it rained for 40 days. As the poles began to reposition themselves, massive changes in weather pattern made tidal waves which flooded out the mountains of the world and as I indicated earlier the book of Genesis tells us 7 times that all people and animals who remained on the land were drowned. Many other texts tell us similar accounts about only those in boats and other vehicles were saved and no pure Cro-Magnon save 8 family members of Noah as it appears that both Methuselah and Lamech, Noah's dad, died in or just before the flood. Here are a few verses to consider.

Enoch 10:17*- And then shall* all the righteous escape, *[the extinction flood] and shall live till they beget thousands of children.* [More than just Noah and 7 family members survived, but Noah's family were the only pure Cro-Magnon. It should be noted that the Essene Jews copied the book of Enoch mare than any other sacred work except for the book of Isaiah.]

Jubilees 6:9- [One of the Anak named "Mesterma" tells God that he should save some of the Anak people-] *and God said Let the* **tenth** *part of them [Anak people]* *remain before him, and let nine parts <u>descend into the place of condemnation</u> And one of us, he commanded that we [the remaining Anak people] should teach Noah all of the medicines.*

Quite a few Survivors

Survivors included 8 Cro-Magnon [Chosen Ones] people [Haplotype K DNA], about 10 percent of the Anak people, and quite a few Gentile humans making up various races with Y-Chromosome DNA Haplotypes as follows. After the flood, they landed at different parts of the world according to 80 thousand similar flood stories.

- C= Negroid landing in <u>southern Africa</u>
- E= Eastern Nubian landing in <u>Central Africa</u>
- G= Proto Europe landing in <u>East Europe</u>
- H= Proto Afghan landing <u>near India</u>
- I = Proto Greek landing in <u>Southern Europe</u>
- N= Proto Russian landing in <u>Mid Russia</u>
- O= Proto Oriental landing in <u>SE Asia</u>
- K= Proto Asian landing in the <u>Armenian Area</u>
- J= Proto Egyptian landing in <u>Northern Egypt</u>
- F= Cro-Magnon landing <u>near Iraq</u>.
- Xx =homo Capensis [sometimes called the Anak] landing in <u>different areas of the Earth</u>.

While I mentioned this before it deserves repeating. This group of mutations made up about 50% of the major mutations of humans noted in the Modern Age with the final group being mutated about 3500 BC. Dr. Joshua Akey, noted geneticist from University of Washington, had this to say-

- *While for a long time there were not many mutations, the <u>human genetic diversity today is vastly different from what 200 generations ago</u>.*
- *A study dating the age of more than 1 million <u>single-letter mutations</u> in the human DNA code reveals that <u>most of these mutations are of recent origin</u>.*
- *Over <u>86 percent of the harmful single nucleotide mutations arose between 5 and 11 thousand years ago</u>. Oddly, since then there have been few mutations at all.*
- *Overall, researchers now believe that <u>about 81 percent of the single-nucleotide variants in the European sampled and 58 percent in the African DNA sampled arose in the past 5,000 years</u>.*

*Of this group only the pure Haplotype F, Cro-Magnon people, were the Chosen Ones who evidently had the **"spirit"** needed to go to the Heaven Universe.*

Religious section

The various Gentiles had a chance to enter Heaven because of the promise that God Incarnate would send his Spirit to strengthen our spirits. All the Nephadim and their offspring and all the Homo-Erectus varieties were all gone and experiencing their own forms of hell. After the flood, some of the Nephilim survived and mated with the gentiles to produce the Anakim race, but both were cursed to become demons who lived for thousands of years in a state of limbo -HELL.

Bharata War Anomaly

Civilization abounded, but soon the lust for power consumed the Nephilim who generally ruled the land. A worldwide war threatened the existence of mankind. The book of Jasher tells us 1/3 of the people of the world died in the wars noted in the Bible and dozens of other texts around the world confirm the horrible war. As we get back to "normal" people we find that everything was fairly good for about 5 thousand years after the beginning of the Holocene. The Anak were teaching science, the various mutated human groups had established cities and infrastructure for reasonable economics and world trade abounded. Many of the Anak ended up in the land of Canaan where the Tower of Babel would be built and construction began around 3500BC of a beautiful citadel. With the aid of the Anak, thousands took part in the building to show strength to the other civilizations of the world. War began during the construction and a massive tower began that would spearhead a new war against God. Nuclear bombs began to drop and those in cities like Mohen-jo-Daro died by the hundreds in the streets while walls were melted. The remains still found radioactive after thousands of years. We don't know exactly when the war started, but we do have a very good idea when it ended.

War Timing Consensus

Sometimes called the <u>Bharata War or the Kurukshetra War</u>, the end of the Babel War is critically timed by the Maya, Egyptians, Hindu and astrophysics as described below.

- **Before 2500 BC-** Zep-Tepi according to *Palermo Stone* from the 5th dynasty Egypt
- **3000 BC-** according to Dr. B.N. Narahari Achar planetary software and astronomical references in *Mahabharata*
- **3066 BC-** according to Dr. D. Abhyankar subtracting 38 war years of *Mahabharata*
- **3067 BC-** from Planetary software and description in the *Raghavan*
- **3090 BC-** Median date for Egyptian *Zep Tepi*
- **3100 BC-** according to Dr. N.S. Rajaram (astronomical statement and interpolated passages of Mahabharata)
- **3100 BC** according to the reunification of the upper and lower Egypt
- **3104 BC-** according to the start of the Age of Kali [Hindu beginning]
- **3114 BC–** according to the start of the Mayan Calendar
- **3127 BC-** according to the *Aihole Inscription* of 7th century AD Egypt)
- **3143 BC-** according to Shri P.V. Holey Astronomical measurements of Mahabharata)
- **3400 BC-** *M*ongulala historical reference- end of the Blood Age [Brazil]

- **3138 BC** -based on astronomy of Saptarishi Mandal the Kurukshetra War dates back to 3138 BC
- **3100 BC-** Excavations in Kurukshetra yielded iron arrow and spearheads dated by Thermoluminence Test to this date

Soon civilizations went underground and we are finding the remains of these underground municipalities all over the world in Scotland, France, Turkey, Malta, Peru, Central America, North America, China, Brazil, Egypt and many other places as new cities are found every few years. While these underground dwellings were horrible to protect from ground forces, they were effective against nuclear devastation. That being said the fallout from the bombing was not as easy to protect against.

Destruction of The Bharata War [3100BC]

Jews called it the Tower of Babel Incident, The Egyptians called the end Zep-Tepi [New Beginning], Indians initiated a new Age of Kali when the thing ended. The Mongulala of Brazil ended their Age of Blood, and the Maya started a new 5000-year calendar. While this was going on, massive mutations formed the remaining races of the world.

Mutations of the Bharata War

For this we will again look at the Male chromosome DNA mutation Haplotypes that mysteriously occurred near the end of the War. The following DNA Haplotype map sort of shows how the populations of the world spread out after the war ended.

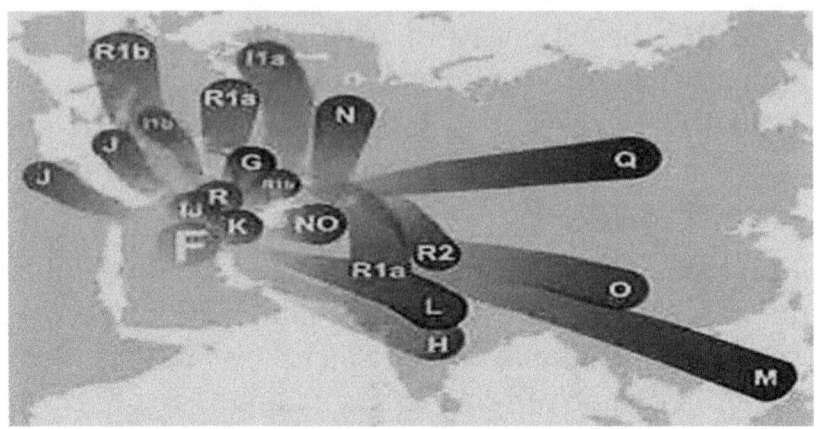

- L-Dravidian India- Dark Indians spilt from "H"
- M- Asian Mongols split from the Haplotype "K"
- P-Proto-Western Asia split from Haplotype "K"
- Q- Northern American Indians split from "F"
- R- Balkan people split from "P"
- R1- Aryan reddish skin colored came from "R"
- R2- Mid European white skin came from "R"
- S-Melanesian Orientals came from "O"
- T-Georgian Middle Easterners came from "G"

Shortened Lives

Other minor mutations followed, but no Major human mutation has been witnessed since this time of destruction and loss of technology. One thing that is very noticeable is how long people live today. The chart following shows the patriarchs of the Cro-Magnon or "Chosen" race. The dotted lines in the chart show the Pleistocene Extinction and another massive war called the Bharata or Babel War. Please notice one thing important in the chart. After the Bharata War, the life span of people became MUCH shorter not during the Pleistocene.

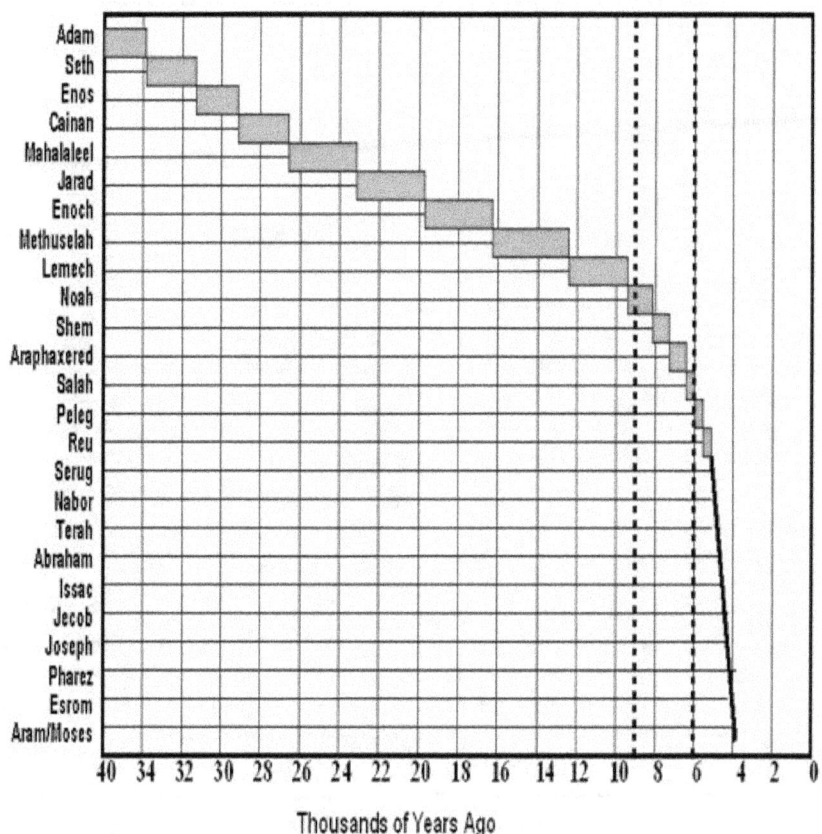

Bharata War Survivors [3100BC]

All over the world we find similar details. The following charts show the Sumerian king lineage [left] and the Assyrian king lineage is shown on the right.

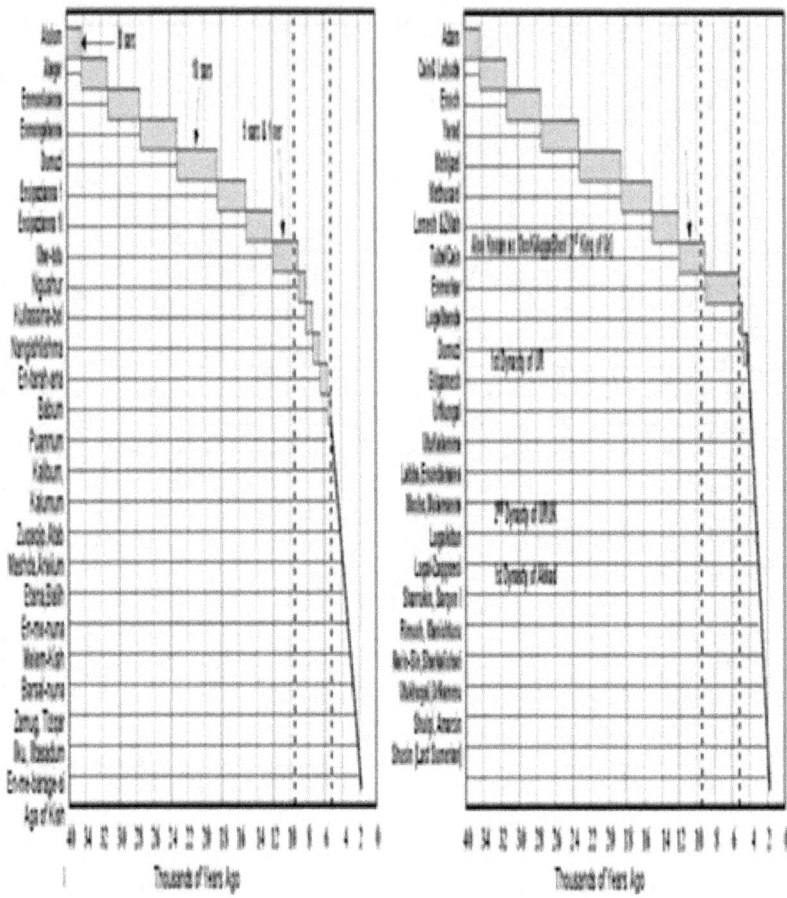

Like the other chart, the 2 doted lines show times of the Pleistocene Extinction and the Bharata War. Finally, we have the king lineage from Egypt that shows the same characteristics.

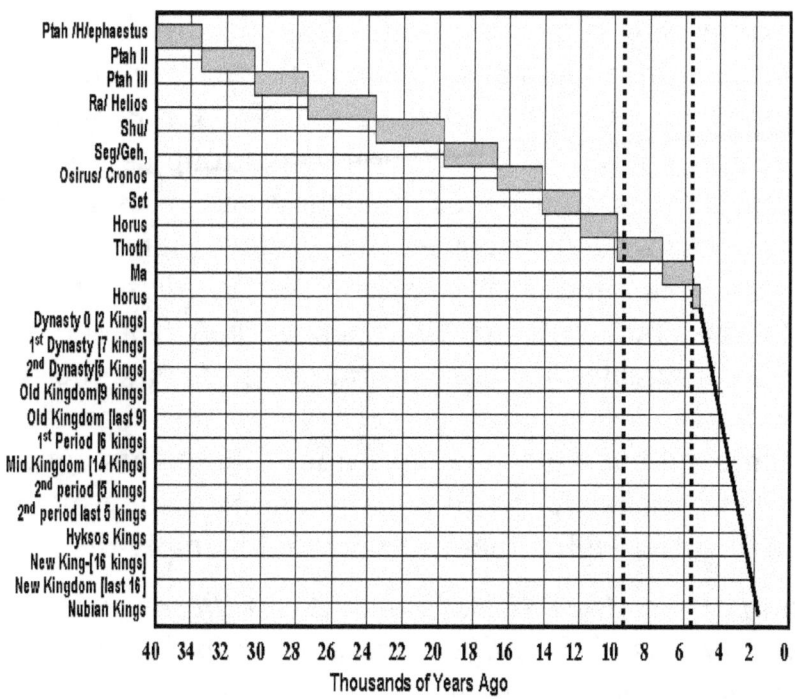

After the Bharata war, something happened to our brains making them atrophy such that modern brains are already 20% smaller than the original Cro-Magnon brains and our lifetimes have been shortened by about 3000%. That is a lot, but some got it worse as they mutated into what we call today Chimpanzees and Bonobos. They became new animals.

Other minor mutations followed, but there were several unusual mutations we should not ignore. In my mind, I would classify the next section as "Hell in Reality" as some mutations were truly horrifying to the unfortunate parents and the unfortunate mutated human.

Ape Mutations [about 3500BC]

Form DNA and Haplotyping we now know the Chimpanzee and Bonobo mutated from humans around the time that the other mutations occurred. Horrible mutations from humans during this terrible time also included another group called the Vanara People. We don't really have to worry about this group as it doesn't affect Heaven and Hell, but they are presented for completeness and because many have ignored these horrors. We find many references to these horrible mutations, but only the Chimpanzee and Bonobo were able to procreate. My book "God Didn't Create the Ape" is one book on this subject I would shamelessly recommend but we can believe not of these gain entrance into Heaven, however, like many animals, they may not have to suffer Hell either as they would simply die.

***Chimpanzee Apes** -Less than 1.5% difference in DNA from humans, 1.5% difference from Bonobo, and 3.5% difference from the Gorilla*
***Bonobo Apes-** Like Chimpanzee, more closely related to human than chimpanzee.*
***Vanara ape-men** of Egypt, India, and SE Asia- Large colonies described protecting human populations with armies, great warriors, and most faithful guards between 3000 and 2000 BC.*

but that was not the worst thing that happened with this technology as a large number of individuals had children that looked almost like apes but the still had some characteristics of humans. This is the group known as Vanara. Let's look at both.

Chimp and Human Differences

Studying Chimp and Neanderthal difference showed reasonable changes so the whole mutation of humans was put aside.

Comparison to Chimpanzee- Researchers compared the Neanderthalis to modern human and chimpanzee sequences. Modern human sequences varied between each other with about 8 substitutions in the DNA chain, but chimpanzee sequences by about 55 substitutions when compared to the mean of modern humans and Neanderthal had about 27 substitutions, so by this 1997 study and another in 2000 we could say Neanderthal was half way between Chimpanzee and Modern man. We will look at chimpanzee later.

Complete Sequencing of Neanderthal DNA Anomaly

Let's look at the DNA charts from before in a little more detail. They seemed to show how very close Neanderthal and modern humans were. This first graph developed in 1999 by Dr. Krings and his DNA scientists shows distributions of sequence <u>differences</u> among Humans, the Neanderthal, and Chimpanzees. This is what they stated.

It may be noted that a small fraction (0.04%) of the inter-human comparisons are larger than the smallest distance (29 substitutions) between the Neanderthal and humans. In this initial work seemed to show Neanderthal could be considered a modern human.

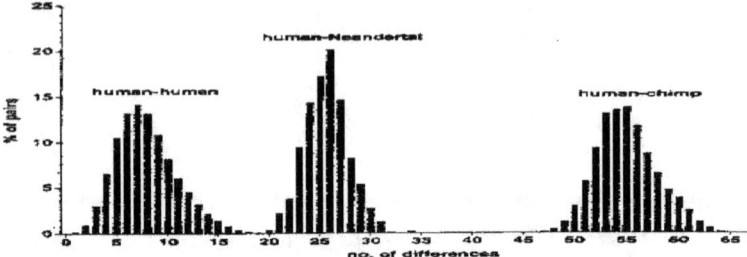

As I said it was found that this was not a correct assumption. In 2008 A team under Dr. Green were able to sequence almost all of the Neanderthal DNA. The chart below is from his study which not only looked at numbers of differences, but also where the traits were positioned in the DNA sequences. It not only shows there are 2 types of modern humans, but also that there was a wider separation than originally was believed. Neanderthal had very small differences to modern humans and much closer than chimpanzee.

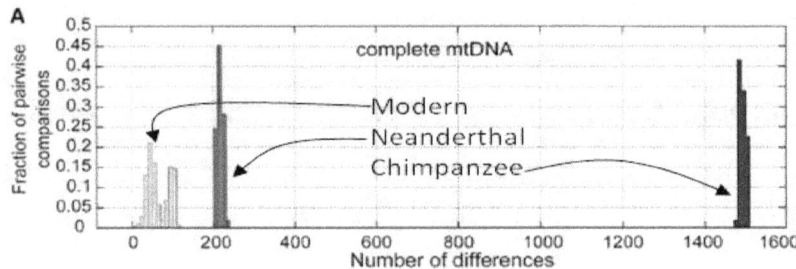

Isn't it strange that Ape DNA has gone through fewer changes than Human DNA? It was as if Man came before the Apes?

Much controversy has been evident today concerning apes and man. It seems scientists are becoming very confused. DNA analysis has shown that Apes may have come into existence after humans, as their DNA has not changed as much as the human DNA. I know that doesn't fit into our neat little world of evolution and survival of the fittest and it may even give concern to us about who should control the world. One might also think that human beings have been around for millions of years rather than the short time assumed by evolutionists. Not only should our understanding be determined by scientific study and conjecture, but also by religious history and cultural legacy. Some may not like bringing in all three of these venues together, but if we don't, then I guarantee we will never have an answer that is remotely close to the truth. The fantastic speed of diversity completely disproved evolution as shown below.

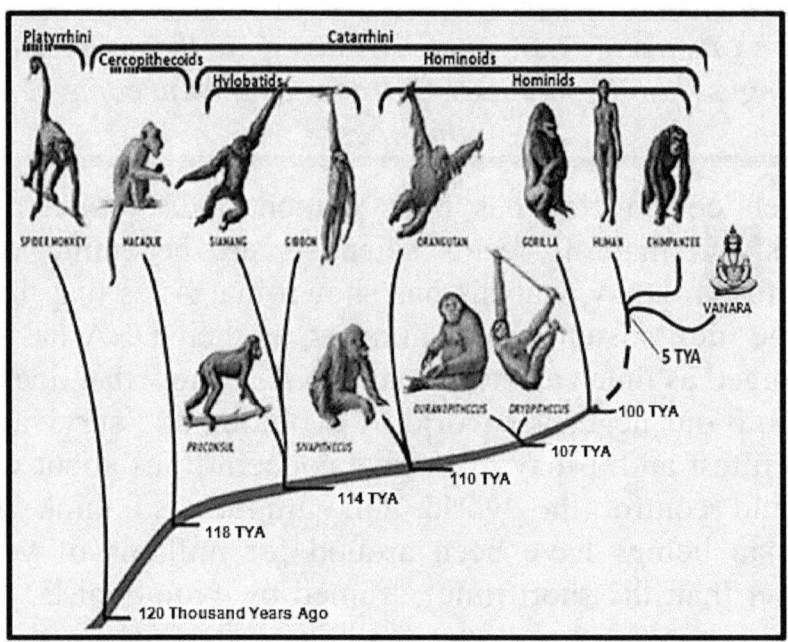

Only Man's Head Evolved-While we are on the subject of chromosomes, the similarities between chimpanzee and human genes tell us a lot about evolution. The chimp and modern man have almost 100% identical chromosome patterns. In fact, the number is greater than 98.5%. This by itself tells us nothing. While that is odd enough, our brain function chromosome patterns are 80% different than those of the chimpanzee. The only way I can figure evolution could have such a marked difference in these two percentages is that only the head was allowed to evolve over some 50 million years while the body was stopped or never started to evolve.

Evolution suggests that only our heads were allowed to evolve while our bodies were stuck in an unevolved state.

This absurdity doesn't make sense and neither does survival of the fittest evolution. You can't tell while you are reading this, but a little tear has welled up in my eye as I think about how the children are being taught the survival of the fittest evolutionary process as a fact of science, when there is no substantial evidence about its truth.

Chimpanzee Mutation

If we want to look at the devastation of man's technologies. This last study should put some things in perspective. To discuss this horrible DNA event that happened during the Bharata War we must include both the Bonobo and Chimpanzee. Some may already know some of this, but Chimpanzee and Bonobo DNA are very similar to humans. Some estimate the similarity to be 98%. The main difference in all apes and humans is that the 2nd set of chromosomes has been split apart as shown in the DNA sequences following. This makes the Chimpanzee and other apes have more chromosomes than a man, but characteristics are still very similar. While all the ape variants may have just been manipulated by the Pleistocene humans, we must also question why the Chimpanzee and Bonobo DNA both show such a short time for mutations and what happened to massive numbers of ape-people that were written about and described around the world. The normal chart is shown below as some common ancestor well after gorilla split from the homo-Sapien line. Some have pushed the separation back 50 thousand years, but there is more evidence indicating the split was only 5 or 6 thousand years ago around the time of the Bharata War.

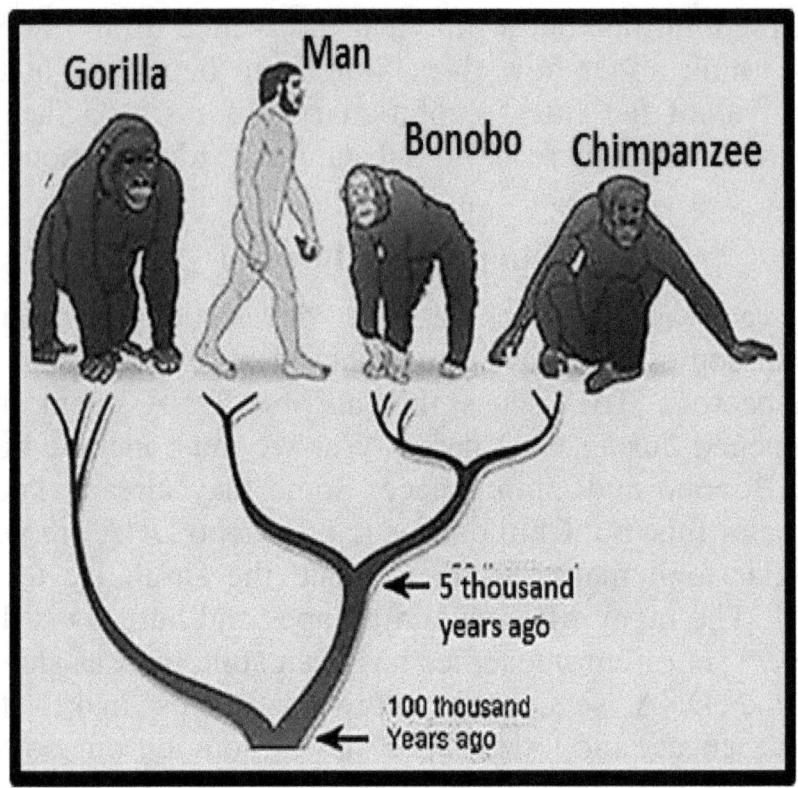

Today biologists are coming around and accepting the possibility of this very recent mutation from their own studies, but many have not connected the Bharata War with the emergence of these 2 primates, bonobo and Chimpanzee.

Ape Mutations [about 3500BC]- From DNA and Haplotyping we now know the Chimpanzee and Bonobo mutated from humans around the time that the other mutations occurred. Horrible mutations from humans during this terrible time also included another group called the Vanara People but only the Chimpanzee and Bonobo were able to procreate.

Chimpanzee Apes *-Less than 1.5% difference in DNA from humans, 1.5% difference from Bonobo, and 3.5% difference from the Gorilla*
Bonobo Apes- *very similar to chimpanzee details but slightly more closely related to humans than chimpanzee.*
Vanara ape-men *we will get to later. Most descriptions are from Egypt, India, and SE Asia- Large colonies described protecting human populations with armies, great warriors, and most faithful guards between 3000 and 2000 BC.*
Homo Sapien Cognatus- *This group may still be a viable human race in some parts of the world. We will look at them after the Vanara sadness.*

Apes After the War-This may seem odd to you but let's investigate anyway. This whole Chimpanzee thing seems like the Gorilla would be the close relative, but it was not. The reason Chimpanzee can use tools so very well is that his closest relative, man, can use them. One possibility is that the brain reducing, DNA changing germ, or whatever it was made some die, some just lost brain capacity some became like apes and a small quantity became chimpanzee. The world had been at war and nuclear weapons had flattened much of the landscape. The Great Tower and citadel of Babel [Baalbek] had been flattened. All of a sudden, chimpanzees appeared. While we're on the subject of chromosomes, the similarities between chimpanzee and human genes, what is important is that, scientists tell us our brain function chromosome patterns are 80% different than those of the chimpanzee. It seems that as the body changes, other characteristics had to change with them. Possibly if we could see a live Homo-Erectus human, we would see a much closer resemblance.

One thing that is very curious is there are twice as many "small" mutations of Chimpanzee DNA over a given time than humans. This suggests the unstable nature of Chimpanzees and possibly other apes as they may all have been manufactured from humans at one time or another.

I'm not spending a lot of time on this subject as it is very important for our understanding of mutation and DNA. The idea that ancient humans used and abused nuclear energy is distasteful to many so no one seems to want to talk about 2 of the worst wars we have ever had on our planet. The first happened about 11 thousand years ago causing about 50% of all major mutations on humans and the second one 5500 years ago caused the other 50%.

Chimpanzee History-The name "chimpanzee" is recorded in *The London Magazine* in 1738, It means "mockman" in Bantu language. Bonobos frequently have sex, sometimes to help prevent and resolve conflicts while Chimpanzees have been seen to barter for sex so prostitution is fairly common. Something we, as humans may have lost during the Bharata War is the fact that Chimpanzee seem to have photographic memories. The locations of Bonobo and Chimpanzee are shown in the following map which also describes places where stories and artifacts of Vanara and Lizard people that we will have to investigate later.

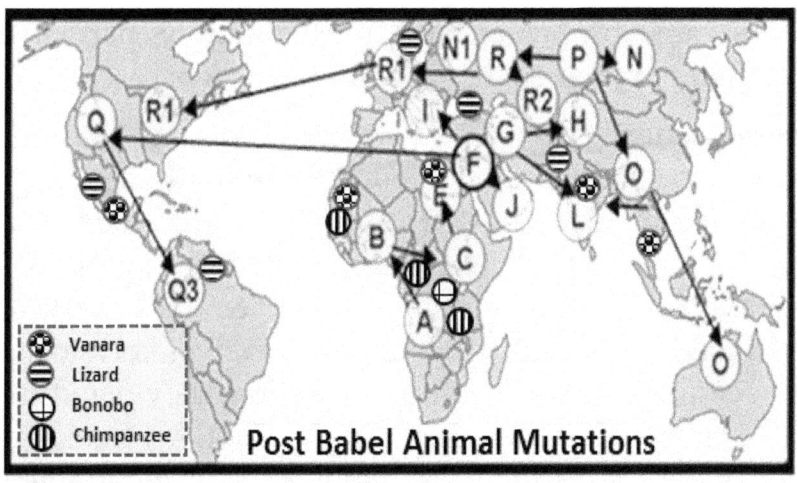

Post Babel Animal Mutations

I know this is not what you learned in school as you were told Chimpanzee and Bonobo evolved separately from man as both separated from the Gorilla lineage about 4 million years ago. ***Today we know this is not truth.***

One thing is for sure, while the evidence keeps mounting that ape-men were useful members of society between 3000 and 2000 BC. Evolutionist scientists backed by consensus rather than fact will continue to lie, disregard, and weave complex reasons for ignoring evidence to hold onto their sacred religion over the ancient Judeo-Christian religious details presented. Besides many documented elements supporting the Vanara and the images, and the statues, and the many leaps of faith required to believe what you were told in school. While we might expect at least one of the Vanara mutations would have been successful in procreation, many scientists don't even want to think about it. Let's test some of the things we are finding out about Chimpanzee and his cousin the Bonobo to see if it makes sense chimps came from man and, most likely as part of the mass mutation that occurred 5500 years ago.

Test number One-A few years ago, one group of researchers studied the genomes of 12 species of Drosophila or fruit fly, four species of nematode worm, and 10 species of primate, including humans. By comparing with other groups of species, they were able to estimate how long ago the genes were likely to have been acquired. Rather than by evolution, they determined that a number of genes, including the ABO blood group gene, were confirmed as having been acquired by vertebrates through intrusion of viruses, protists, fungi, and Bacteria. They confirmed 145 genes were acquired by this means to shape humans. <u>They found that 50 additional genes were "donated" in chimpanzee</u>.

Test number Two-DNA structure of Chimpanzee is almost a complete match with humans. While one of the chromosome strings has been split in Chimpanzee, the makeup of the DNA sugar is very close. The following image is of a Human, mouse, and chimp X chromosome containing about 1,100 different genes, or sets of instructions. Each gene affects a particular trait in the body. Each specific nucleotide that makes up a DNA string [adenine (A), thymine (T), guanine (G) and cytosine (C)] show up as a slightly different shade in the image below. Notice that the Chimpanzee and Human are virtually identical. This is very strange. You can see how different even the mouse DNA is. Even the Centromere [necked down area] is the same in chimps and men.

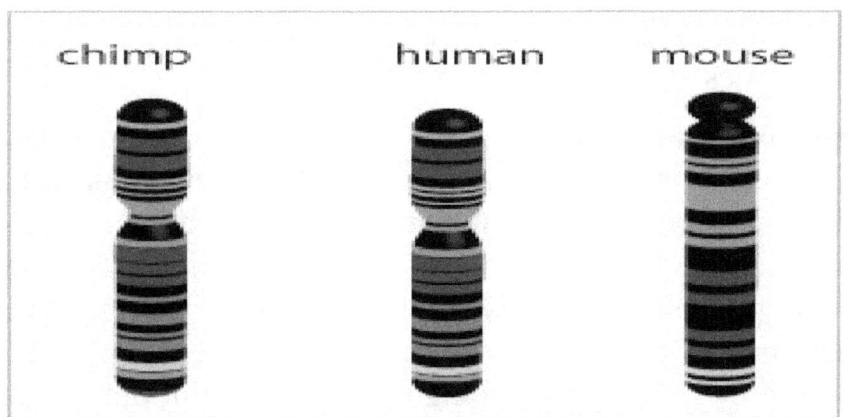

As I showed before just looking at DNA sequences of Orangutan, Gorilla, Chimpanzee and Modern human shows chimps and humans are especially close as most to the proteins line up exactly in line.

Test number Three-Human and chimp DNA was determined to be 1.2 percent different. Gorilla and other apes have over twice [3.1%] that much difference as they are somehow very different. In should be noted the genetic difference between individual humans today is minuscule – about 0.1%, on average. Now for the really weird part; bonobo has about 1.2% differences like chimps but Bonobo and Chimps have 1.6% difference between their DNA and Gorilla and other Apes.

Chimps and Bonobo are more closely related to humans than apes or even each other.

Test Number Four- At the end of each chromosome is a string of repeating DNA sequences called a telomere. Chimpanzees have about 23,000 base pairs of DNA that are repeated. While humans only have 10,000 base pairs of DNA repeats. One could determine that Chimps have not gone through as many mutations collecting these duplicates and is a newer species.

Test Number Five-It was determined that Bonobo was a mutated branch from Chimpanzee that occurred about a million years after the Chimpanzee and Human split 4 million years ago [using nuclear decay timing]. This has been augmented recently. In three separate studies, it was determined that the human chimp split could have been only <u>6500 years ago</u> making the Bonobo split only about <u>5 thousand years ago</u>. [In 2001"*Phylogenetic And Familial Estimates Of Mitochondrial Substitution Rates: Study Of Control Region Mutation In Deep-Rooting Pedigrees*"; 1997. "*A High Observed Substitution Rate In The Human Mitochondrial DNA Control Region*". In 2000, "*The Mutation Rate In The Human MtDNA Control Region.*"]

Test Number Six-The following chart comes from the journal "*Nature*" and has nothing to do with me or the seemingly crazy details I am trying to tell you about. The reason I put it here is I wanted you to look at where they placed the "appearance" of chimpanzee up in the right-hand corner around the same time as modern man. While they wanted to show commonly discussed species, they did not try to determine hypothesized lines of descent, just when we believe they "appeared". Disregard the times as I discussed previously as this is using nuclear decay timing still.

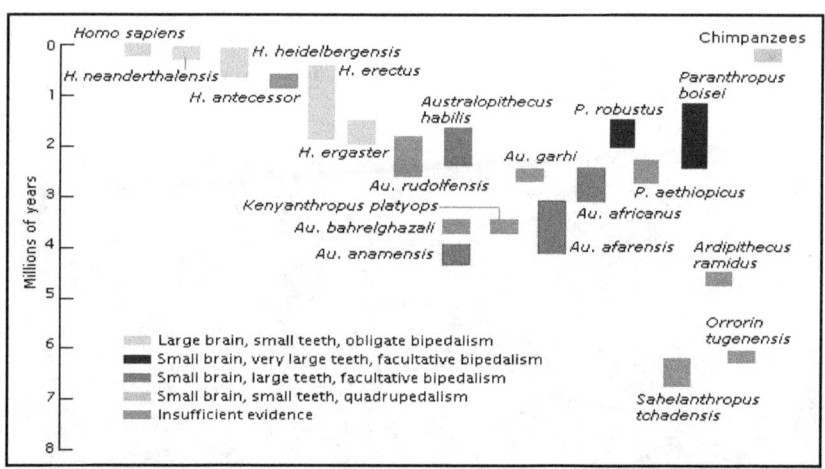

Test Number Seven-In the study, *"Bonobo Genome Compared With The Chimpanzee And Human Genomes"* Dr. Eichler and his colleagues found that the human and chimp sequences differ by only 1.2 percent in terms of single-nucleotide changes to the genetic code, but 2.7 percent of the genetic difference between humans and chimps are duplications, so we could really say it's only 1.1% difference. They also found that more than <u>3% of the human genome is more closely related to either the bonobo or the chimpanzee genome than these are to each other</u>. They also found almost a thousand integrations of transposons [Transposed similar sequences] absent from the orangutan but present in bonobo, chimpanzee, and human. Of these, 27 are shared between the bonobo and human genomes but are absent from the chimpanzee genome, and 30 are shared between the chimpanzee and human genomes but are absent from the bonobo genome. The images below are of the Bonobo, Homo-Erectus, and Chimpanzee, showing what the small changes do to a person. In addition, about 25% of human genes contain "parts" that are more closely related to one of the two apes than the other. This suggests that Bonobo did not necessarily split from Chimpanzee. It could have been a new mutation of Human.

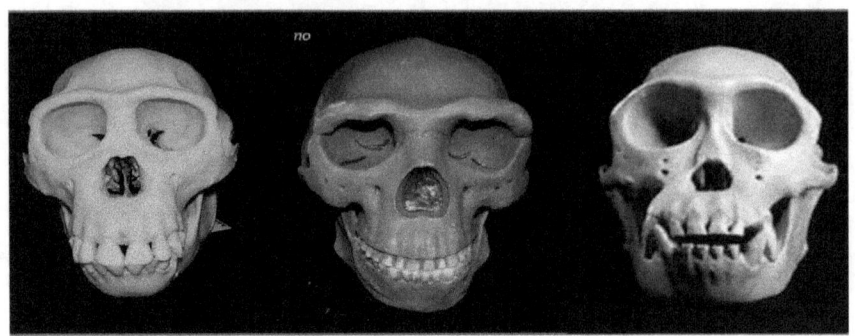

The following image compares Australopithicus, Erectus, Modern man, Bonobo and Chimpanzee skeletons. While the hips and hands have reverted back to Austropithicine style, there is a lot of similarity to the bonobo and chimp to homo-erectus. The bonobo is especially similar as noted from the skull and teeth similarity shown previously.

Science Confirms the Horrible Mutations-Before you start thinking I'm a nutcase, let me show you a study by Dr. McCarthy and noted Geneticist. He wrote the following: *"The average heterozygosity of chimpanzees is less than one-fifth that of humans. --Heterozygosity is a commonly used measure of genetic variability. If humans and chimpanzees were descended from a single ancestral population, mutation would have greatly increased the level of human nuclear variability relative to that of chimpanzees during the time since the two diverged.* [Humans have been around 5 times as long as Chimpanzees] *The actual observation, however, is that chimpanzees have "three to ten times" the mitochondrial* *variability of humans.* [This means the DNA is much less stable as if they mutated from humans.]

Since Cro-Magnon appeared about 30 thousand years ago, this would mean Chimpanzees mutated about 6 thousand years ago.

The following images are of the various major variations in humans over the years. Please notice how very similar the Bonobo and modern man skeletons are. Yes, both bonobo and chimpanzee went back to the australopithecine hands on the feet style, but should the Vanara and those that mutated farther be allowed in Heaven?

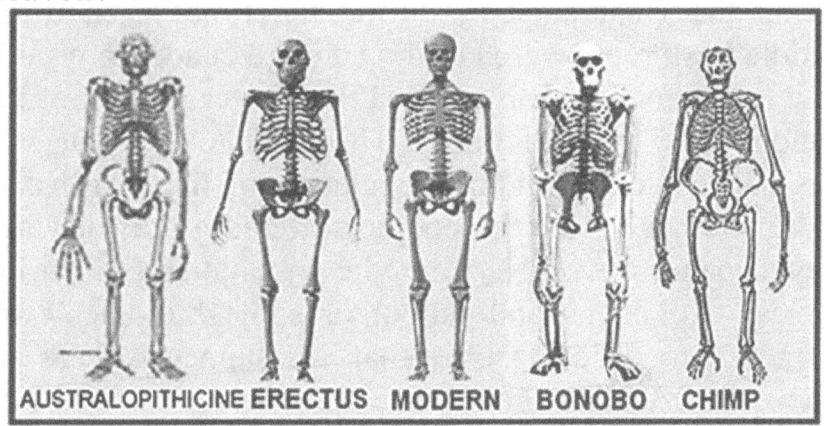

Hopefully you are beginning to understand used how horrible our technology can become if we don't try to control it. I was going to put in some of the "mistakes" our DNA splicing scientists have made just fooling around, but instead I think a quick look at the horrors of becoming Half-Ape.

Vanara Mutation

It would be an embarrassment to say ape-people lived 4 thousand years ago. It would be better to just disregard the hundreds of descriptions, images, and other evidence.

For this discussion, let's look at the holy book of Jasher. This gives us the mutations and the timing as being during the Bharata War, during the time when many Middle Easterners were building a huge citadel in what is now Lebanon. The Bible called the Tower portion of the Citadel the Tower of Babel. The book tells us the war was harsh and 1/3 of the population of the world died while another 1/3 of the people suffered a horrible fate, they became devolved from the mutations felt during nuclear explosions and fallout or some biological war germ that modified DNA. Whatever happened, here are the words from this ancient Jewish writer.

Jasher 9:34-36*- those [Babel Tower Builders] who said, "We will ascend to heaven and serve our gods",* <u>*became like apes and elephants*</u>*.* [The interpretation is that some became like apes or ape-men and something even worse about 3100BC.]

Around the world we find the exact story. Here are a few from the Aztec, Maya, Totonac, Lower Congo Tribes, Tibetans, Egyptians, Indians, and the ancient Chinese. Luckily or unluckily, the DNA of this group was not stable enough to stay long in our history, but we should never forget them.

Here are some of the descriptive texts from around the world.

Mahabharata - Like Ramayana, this set of sacred Hindu books describes the Vanara ape-people as forest dwellers, and mentions 2 of their kings doing battle with and being defeated by a Pandava general who led a military campaign to south India. It also makes a big deal about Hanuman as being from the gods [Homo Capensis people] of that time. To save Princess Sita, Hanuman leaped the distance between India and Sri Lanka to save her. Another story has Hanuman going through a tunnel system to a new world. [and similar ape-man stories are found there as well.]

Totonac- Mexican Tradition*-After the flood, the boat carrying the survivors finally rested. <u>God reversed man's face and hind parts. If that wasn't enough, he then turned him into a monkey</u>.* [Probable indication of man turning to apes or ape-men after the Babel incident]

Mayan Tradition*-During the second creation, <u>people were turned into monkeys</u> and the world was destroyed by wind.* [Possibly talking about the destruction of Babel and another indication of people turning into apes or ape-men]

Aztec History*-During the Age of the Four Winds <u>men were turned into monkeys</u> according to Codex "Laticano-Vatino"* [We are told wind destroyed the tower of Babel and the ape or ape-men thing keeps coming back.]

***The Lower Congo Tradition*-**First God created man. Then, after a huge flood, <u>men put their "milk stick" behind them and were turned into monkeys</u>. [Some men became primitive like apes or ape-men after the flood as was indicated after the Babel incident]

Tibetan History*-Tibet was almost totally inundated by the flood [*Pleistocene Extinction tidal waves]. The <u>survivors had been little better than monkeys,</u> until the god Gya sent teachers to civilize the people and they repopulated the land after the flood.*

Egyptian Text-Emerald Tablets*-The master said-take them by the arts ye have learned of far across the waters until ye reach the land of the <u>hairy barbarians</u>, dwelling in caves of the desert. Follow there the plan.* [possibly Vanara People]

Chinese Ape men-From "Journey to the West", we find SUN-WU-KONG: [The Monkey King, Trickster God, and Great Sage Equal Of Heaven]. Here is a short sampling of description of the revered apelike human. *"From the beginning of time, a certain rock on the Mountain of Fruit and Flowers had been soaking up the goodness of nature and QI energy. One day this pregnant rock released a stone egg, and from it hatched a Stone Ape, who solemnly bowed to the Four Corners of the Earth— then jumped off to have fun. This was SUN-WU-KONG. He was high-spirited, egotistical and full of mischievous pranks. He was soon having a wonderful time <u>as King of the Apes</u>. But a niggling worry began to gnaw at him — one which would change his life. The <u>Monkey King feared Death</u>. To find immortality, SUN-WU-KONG became the disciple of Father Subodhi, a* *rather dour Daoist sage. The sage, unimpressed with his simian tricks, gave the <u>Monkey King a new title: 'Disciple Aware of Emptiness'</u>. SUN-WU-KONG was very pleased with this epithet, not realizing it referred to the vacuum in his head."*

Indian History-India has a rich history filled with ape-men during the time of the Bharata War. Their 2-main religious historical works, Ramayana and Mahabharata are filled with the descriptions and events associated with the devolved people known as the Vanara.

Ramayana 17:8-18*- Vanaras [ape-people] are created by Brahma to help Rama in battle against Ravana. They are powerful and have many godly traits. Taking Brahma's orders, the gods began to parent sons in the semblance of monkeys. They are powerful and have many godly traits.* It presents them as humans with reference to their speech, clothing, habitations, funerals, consecrations and describes their monkey-like characteristics such as their leaping, hair, fur and a tail. Some were turned in anger. *One of the gods* [Homo Capensis] *Gautama, had a daughter named Anjana. Anjana told her father about Indra visiting his wife while he was away. So, his wife cursed her daughter to turn into a monkey. Gautama cursed his two boys to turn into monkeys as well as they had failed to inform him of the same. He could not retract the curse and so he gave all three monkeys to Riksha, the monkey-king of Kishkindha. anaras are created by Brahma to help Rama in battle against Ravana. ----After Vanaras were created they began to organize into armies and spread across the forests, although some, including Hanuman, stayed with*

"Normal" humans. The evil king of Sri Lanka, Ravana, kidnapped Prince Rama's wife Sita. He and the Vanara army led by Hanuman battle against Ravana. Two of the Vanara, Nala and Nila, built a bridge over the ocean so that Rama and the army could cross to Sri Lanka, kill Ravana, save Sita, and bring peace to all of India.

There is absolutely no question that without the help of the Vanara [Ape-people] The Indian Civilization would be very different today and substantially more evil. Therefore, their images are depicted just about everywhere and their religion describes their plight. Some of the hundreds of images are showing in the previous collage. In Egypt we find similar trust, and aid by these people and in Cambodia, the Vanara has places of high esteem in society. Just as many images have been made to describe how "loved and trusted" the Vanara were found in many places.

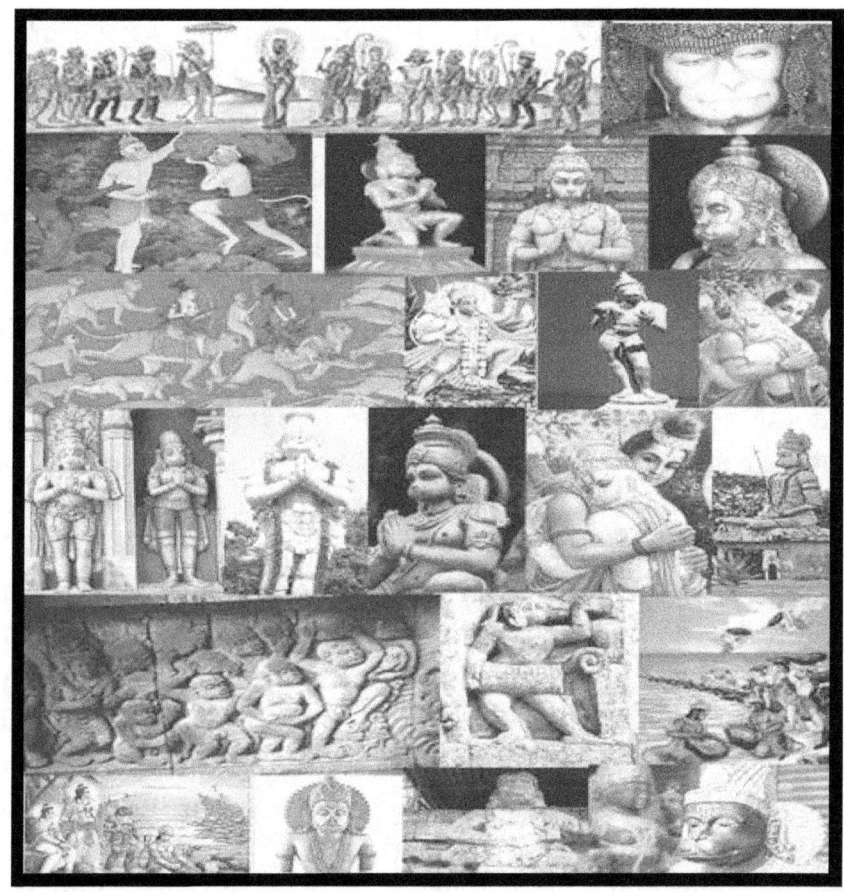

Mahabharata- Like Ramayana, this set of sacred Hindu books describes the Vanara ape-people as forest dwellers, and mentions 2 of their kings doing battle with and being defeated by a Pandava general who led a military campaign to south India. It also makes a big deal about Hanuman as being from the gods [Homo Capensis people] of that time. To save Princess Sita, Hanuman leaped the distance between India and Sri Lanka to save her. Another story has Hanuman going through a tunnel system to a new world. [and similar ape-man stories are found there as well.]

Let me just reiterate the claim of people being turned into monkeys and apes. One of the horrors of the 3rd Armageddon was this negative evolution of man to assume the nature and appearance of Apes. The book of Jasher actually indicates that 1/3 of the total world's population became like Apes and Elephants, but let's not worry about elephants until we get through with the Apes. Besides, there are so many texts referring to this APE horror that it is an important description to understand. What we know from physical evidence is that just about everywhere, apes and monkeys somehow became important, civilized members of societies. How odd is that? Some try to ignore this fact, but it doesn't change anything. Monkey, ape, monkey, or ape---you should decide on your own. If there were only a few, that would be one thing, but the facts seem to indicate what we are told in the book of Jasher. While some of the people of the world lost their memories and ability to communicate without speaking, others got a bigger dose of whatever it was and they reverted, physically, to an apelike character. They were still men but they had the appearance of an ape. The people that recorded this startling event from the war include the people of the Congo, the Totonacs, the Aztecs, and quite a few others. All of these people indicated that some of the people became monkey-like some time after the flood. You will see from the data that this theory is not so unreasonable. Whenever many groups of people say the same thing, the probability that it is the truth becomes more likely. Let's read the histories so we can test probability.

Cambodian Vanara

In Cambodia, we find the same thing as the Vanara were so revered they were described as the protectors of the famous Angkor Wat temple.

Around the country more and more images of vanara can be noticed.

African Nomoli/ Vanara

There is no telling how many places the Vanara lived in around the world but we can believe ape-men who had been part of the Western African society lived during the time of the Vanara. This time they would be called the Nomoli. Today there is not much in Sierra Leon, but there are diamonds so many, search for diamonds. Instead of diamonds they kept finding these Nomoli statues. No one knew why they were on the ground characterized as partially apes, they mostly squatted, had huge round ears and the muzzle of an ape, but they were also human-like very similar to the Vanara described in India. One legend in the area says: *The part of the sky in which the Nomoli lived turned to stone. It splintered and fell to Earth as pieces of rock.* Another legend indicated that *angels had once lived in the Heavens. One day, as a punishment for causing bad behavior, God turned the angels into humans and sent them to Earth. The Nomoli figures serve as representations of those figures, and as a reminder of how they were banished from the Heavens and sent to Earth to live as humans.* Another legend indicated that *the statues represent the former kings and chiefs of the Sierra Leone region.* There were indications that an ancient group called the Temne would perform ceremonies during which they would <u>treat the figures as if they were the ancient ape-like leaders</u>.

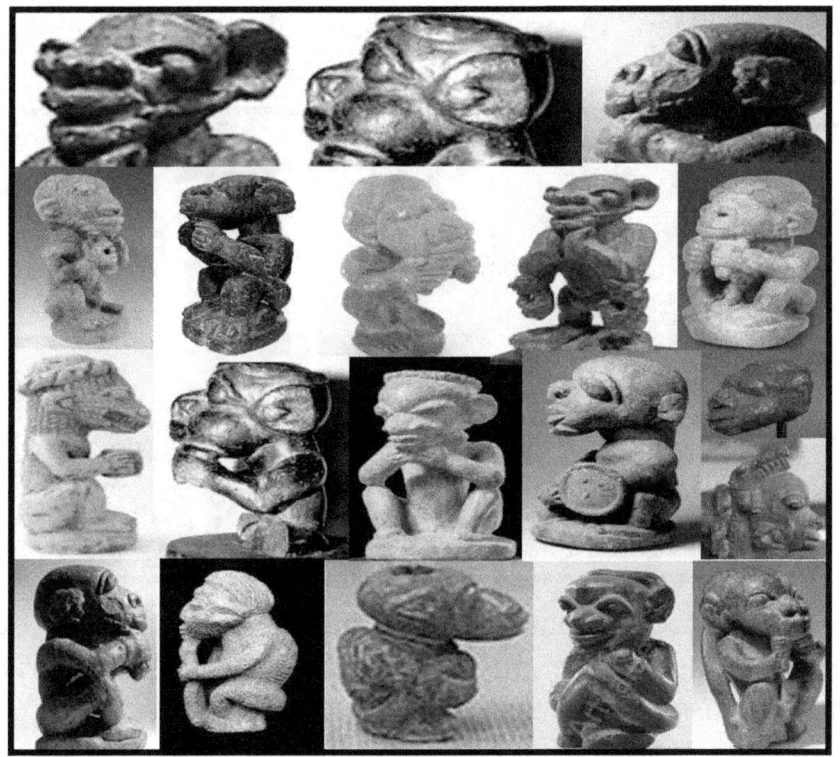

The previous collage is a small sampling of the huge number of these figurines reminding us that some became like apes as the book of Jasher told us. We even know the statues were made around the end of the Bharata War, before the eventual loss of DNA messaging which caused memory and capability loss. When one of the Nomoli statues was cut open, a small, perfectly spherical metal ball made out of steel and chromium fell out which we only now are able to process.

In the Americas- We find the same type of ape-like people. The 2 on the left and middle top are from various Maya. The middle bottom is from Mexico and the Aztec. The top right image is from Guatemala from the City of Monkeys, and the bottom image is from Panama. All show a deep reverence for ape-like people in the Americas.

While it seems that the Nomoli or Vanara died out centuries ago, there might be a cousin that did survive these thousands of years.

Where Are the Ape-men Now?

Possibly there were laws against intermarriage, but generally speaking the ape-people had jobs and held positions of authority. As the change was a DNA change, we can believe their offspring would have either amplified the characteristics or made the DNA structures more unstable. We know that chimpanzees, for instance have twice as many mutations in their DNA as humans attesting to the fact that their DNA is more unstable. With such a major change in DNA, we can believe marriages between normal and ape-people would not have successful offspring and the mutation rates of their offspring would have been amplified as the DNA bonding might have been more unstable than sequencing grounded by ancient development.

Hopefully, you are getting a stronger belief that there might have been truth in the many different and generally isolated writings. I know it still sounds absurd to you, but just about every society is telling us that it did happen and there is more evidence to come. If we assume for a minute that hundreds of ancient people were not completely wrong in their observations, it would be interesting to trace the histories of these unfortunate people. All we can assume at this time is that a large group of humans had some of their "human" genes removed leaving more of the hairy ape-man characteristics like Homo-Erectus man. The offspring of these "punished" individuals possibly became hairier, and even had a more ape-like appearance just like the Homo Erectus had previously.

The ancient references do not indicate if these humans were transported to another location as was done with the other 1/3 of the population, so we can possibly assume that many stayed in a Middle Eastern location. This group must have quickly inbred others with similar characteristics and regained a place in our society. We can assume they had limited procreative capability and continuation of their species must have been a very hard road. Finally, the instability of their DNA would not allow for breeding and they were lost to history. Whatever happened, the inbreeding has now completely removed this "special group of APE-humans", but at one time there may have been "apelike" humans and that time wasn't millions of years ago. Possibly there still are throwbacks.

Modern Monkey Tails

While tails are generally a sign of a monkey rather than an ape, for the purposes of this book all primates outside homo-sapiens are put into the same category. In India, the monkey tail is back. The first image below is a boy with an actual tail revered by the Hindu zealots as some type of their incarnate hero. The Ape-man, hero Hanuman has returned! The boy just thinks it's something to sit on as the ape tail is beginning to reappear. The question might be, "Are some monkey or ape characteristics returning?"

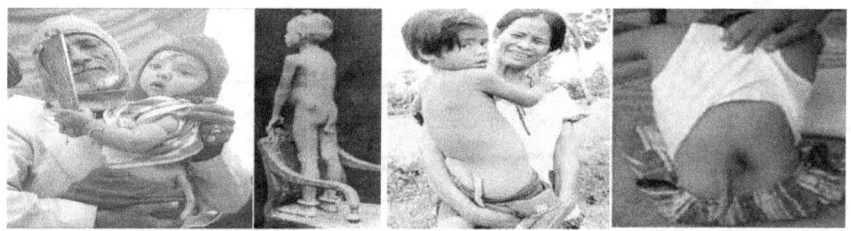

We are seeing more and more monkey tails reemerging as well as other scary features. While we all start off with a monkey tail as shown below, many times the tail is lost in the development process.

 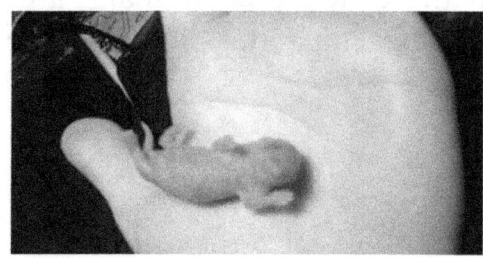

Unfortunately, more and more people are retaining the tail possibly as DNA combinations are just matching up or there may be some mutation bringing us closer to apes again. Here are a few more examples.

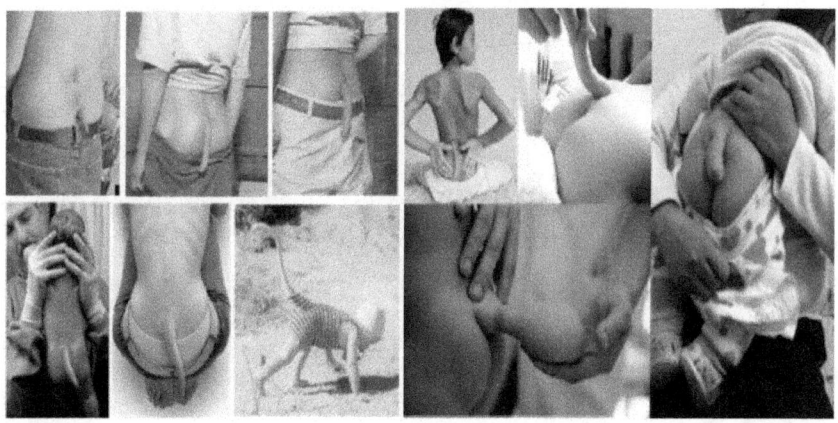

Year after year the quantity of people with this new characteristic seems to be increasing which suggests a reemergence of the features of the post Babel Ape-men. Who knows? Besides the disfiguring mutations I've been addressing, there were more serious ones that were done either intentionally or from nuclear fallout, or from some biologic agent, but we can be pretty sure, two other massive mutations occurred besides the Vanara devastation. Those were the Chimpanzee and Bonobo Mutations. Unlike the Vanara, the last two mutations were even more severe but they allowed for procreation so these devolved humans are still around. DNA can help us get a better picture of the Homo Sapien Cognatus.

Homo Sapiens Cognatus

This anomaly is fairly well known. Cognatus is just a fancy way of saying Big Foot. I'm sure you are thinking I'm plucking a banjo right now and chewing tobacco, but before we discard this strange mutation and just identifying these humanoids as anomalies, we need to do some investigation as they have been seen around the world and thousands of sightings in the United States continue to increase as this elusive and rejected, human-like-ape tries to stay hidden. The images below generally describe the humanoid that, like the earlier Vanara, have been seen in almost every major country of the entire world.

 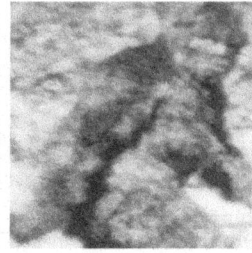 

There has been a concerted effort to classify the DNA of this "creature" over the past 10 years. A 5-year study of 111 samples of hair, blood, and scrapings tested in 34 different laboratories, has taken the DNA samplings to classify this animal. Of the 111 a number of the samples were found to be cat, dog or other animal but a large number were classified as human-like. In fact, the Mitochondrial DNA was determined to be human with 5-nucleotides markedly different than "normal" people and when chimpanzee DNA was compared, they found a 5-

nucleotide difference as well. If I didn't tell you already, the reason to use Mitochondrial DNA is that mtDNA genes have a much higher mutation rate than the coding regions of nuclear or Y-Chromosome DNA genes. They found that all three major mutations defined as haplotype "T" were present [Middle East/Turkey], but none of the mutations associated with "T2" haplotype [Middle Europe] describes when the separation of species occurred. As we would expect, like the Vanara, Cognatus mutated from humans about 6 thousand years ago. Interestingly the Cherokee Indians have a very similar haplotyping. In no way am I saying Big Foot humans are Cherokee, it's just an oddity.

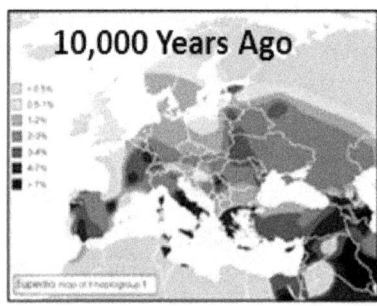 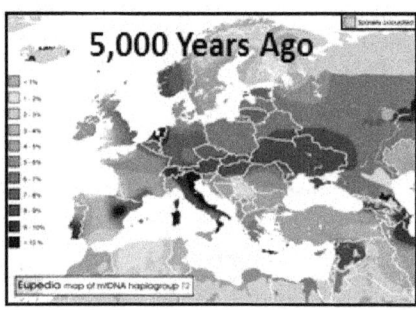

I'll bet that was a shock. Here is another one. Now that scientists know this is a mutation of a human, you would think the information would be placed in our biology and anthropology textbooks. Just kidding! That would give children too much information. The map following shows just some of the sightings around the country. There can be little doubt that the remnants of the Vanara people turned into ape-men after the Bharata War are still around and abound in the United States and around the world.

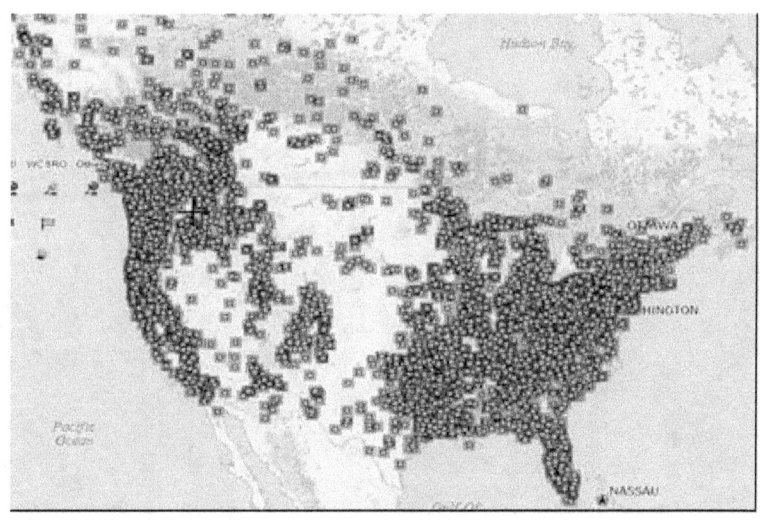

While these people are elusive, they have to walk so sometimes, conditions are right to see footprints. At about twice the foot compression area as we find from normal people, we find that the feet have a wider appearance when compared to smaller footed people.

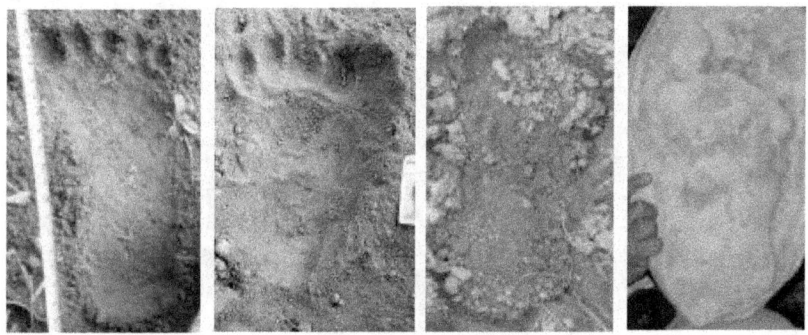

Every year more and more of these hominoids show up and more and more video as something appears to run away from intruders. Where the images were taken, footprints are left.

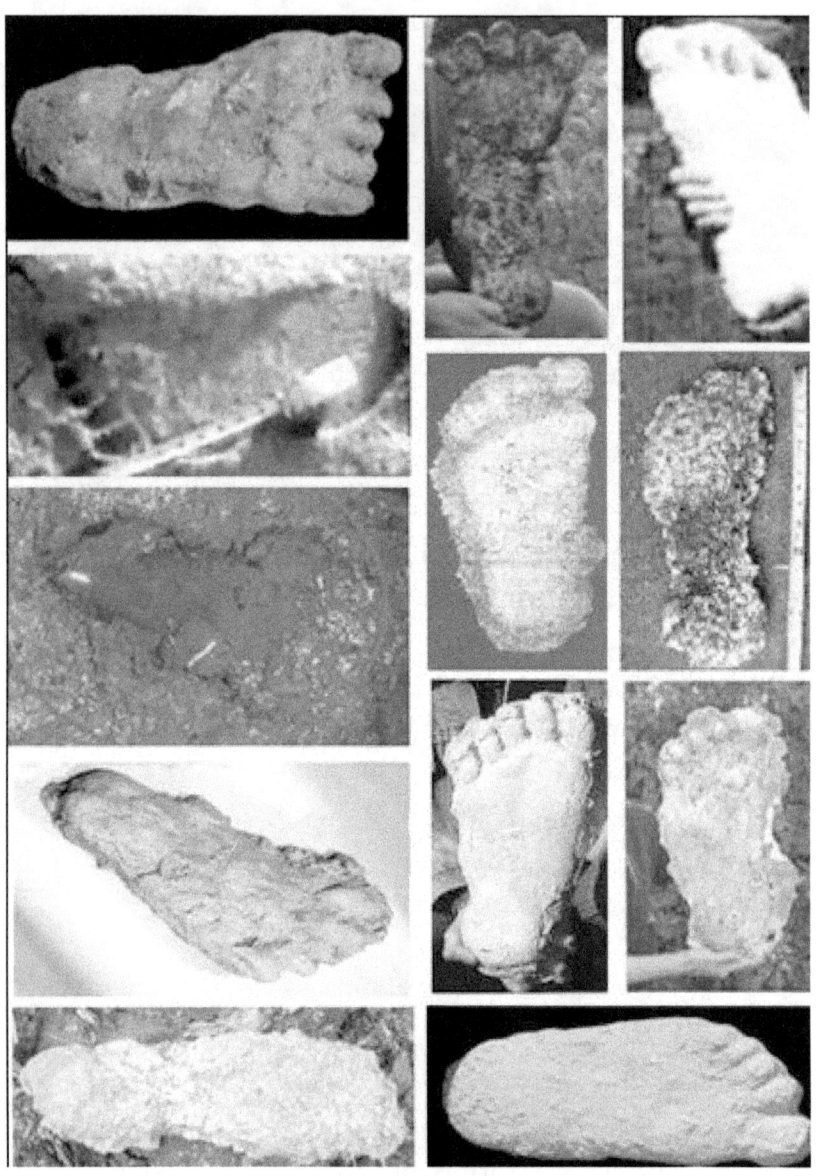

Let me show you a slight difference in our 2 races. Our hair is much finer and is not scaly. The hair on top is normal human and the bottom one is Sasquatch, I mean Homo Sapien Cognatus. There is no reason to make them feel any worse that they already feel. The first shows a clump of hair found. The second image shows the difference in hair samples; and the last image is a close-up of the Cognatus hair.

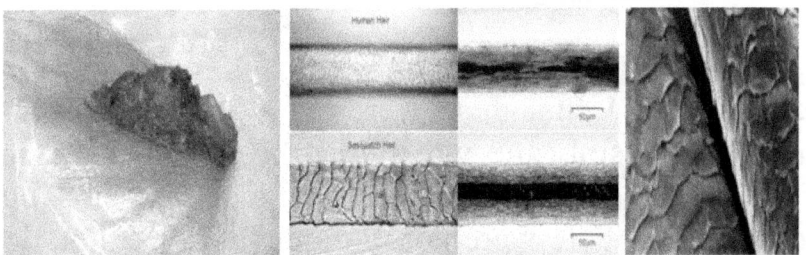

I know some of you are still considering Sasquatch as an fake anomaly, so let me go on to a human mutation that was just as sad or worse.

Ubaid People Anomaly

According to Jasher there were many tragedies in DNA mutation, Possibly the worst was the Jasher description of some people devolving to become like elephants, but it did not mean there were not more devastating mutations. While not made like elephants, it seems one major group of people mutated to become leathery like an elephant, possibly, but they looked more like lizards to me. I probably don't need to go through this section as they were so widespread, the descriptions should be in every anthropological textbook. Never mind. I forgot, what we can laughingly call scientists simply say these thousands of unfortunate people were anomalies. Let me first say there are variations in this group that are more significant than the Vanara, but there are many similarities. The initial finds of statues of these unfortunate people came from the preSumerian group called the Ubaids, so all lizard people, sort of, gat that name by default. We do not know what the lizard people were actually called, but we can tell they were welcomed into society immediately following the Bharata War. First let's read what the ancient said about them besides the Jasher comment.

Sumerian Text- *Isis kneaded it in her hand with some dust and she <u>fashioned it in the form of a serpent.</u>* [Just like the many similar stories, we may believe that "the gods" turned some humans into snakelike beings.]

Babylonian Text- <u>Edimmu</u>, *were a type of* <u>viper-faced demon</u> *considered vengeful toward the living. They sucked the life out of the susceptible and the sleeping children and young adults. Most common symptoms were lightheadedness and the sensations or being groped.* [We can believe that any description of demon was one of ancient people who died and because of some special capabilities while alive, many believed they would have these capabilities after death. The only thing we care about here is there once <u>was viper-faced people.</u>]

Jewish Gnostic- "Hypostasis of Archeon"- *Then the female spiritual principle became a serpent.*

***Jewish Essene "Testament of Amram"**- One of them was <u>terrifying, like a serpent.</u> He was many-colored and dark. <u>His visage was like a viper</u>*

Indian Mahabharata-The Nagas were reptilian beings with snake heads. *The sage Kasyapa had two wives, Kadru and Vinata. Kadru laid 1000 eggs which hatched into snakes, and Vinata laid two, which hatched into the charioteer of Surya the sun god and Garuda. Garuda was later captured by the cousin snakes and had to relinquish the elixir of immortality. Afterward, snakes and he remained enemies.* [We can believe the Naga had been devolved as the book of Jasher indicated.]

Maya-The books of the Mayans called Chilam Balaam say *the first settlers of the Yucatan in Mexico were* **<u>the Chanes</u>** *or <u>"People of the Serpent".</u> They were said to have come across the sea led by a god-figure called Itzamna, a name that <u>apparently comes from the word itzem, which translates as "reptile". Itzamna, the sacred city of the god, therefore, means "city of the lizard".</u>*

Coffee Bean Eyed

We find hundreds of examples of this type of devolved human. The most common of the images of lizard people are the coffee bean eye ones. This is especially true in ancient Sumeria and Babylonia. All over the world we find these images of the same type of person so we had better take a second look. These images [See next left] are from Mexico, Egypt, Sumeria, Kosovo, and China, but there are many, many more where these came from. These were people that lived in society and worked with the people. Just because we don't like the way they looked does not mean we have the right to eliminate their existence from our minds and histories. These were not Gods, but they certainly had a difference about them. Examples are shown below left.

 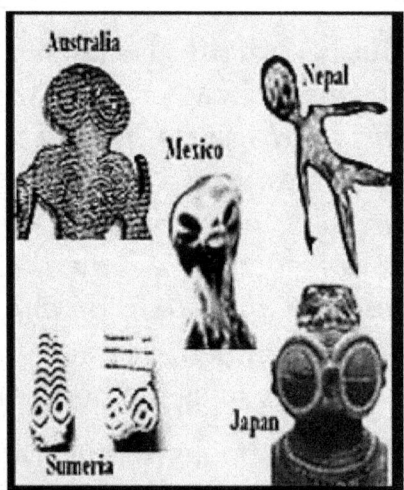

Big Eyed Lizards

Another group could be known as the big-eyed lizard people. Again, we find that there are dozens and dozens of images of this reptilian looking group of people all over the world from 4000 years ago and more. In Japan, this group is commonly **known as Dogu**. I don't know what it means, but there is no mistake that these people looked very different than the other "normal" people of the day. [See above right.]

Pointy Head Lizards

Like the other Lizard People, evidence of the pointy-head group can be found everywhere. A small sampling is shown below. There can be little doubt that this group did not fit in with the normal people, but somehow, they assimilated into the population. The fact that they are remembered in effigy shows that they were not only a part of the society, but they were an important part and somehow, we forgot about them. Generally, when the eyes are visible, they are located on the sides of the face and all they got for a nose was 2 nostrils. These were not normal people. Their face looked more like a dinosaur. See below.

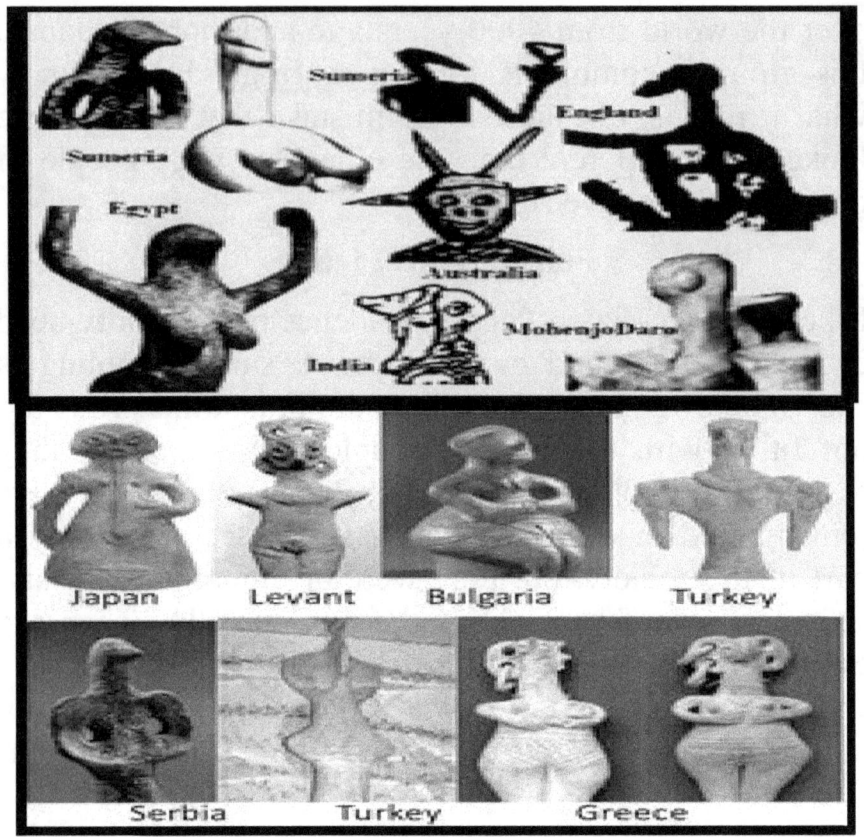

Snake -like People

While many of the lizard people that lived during the ancient times had human bodies and lizard heads, there can be little doubt that a similar group of people shared some common ancestry. Like the others, this group integrated into life with normal people around the world. See images following.

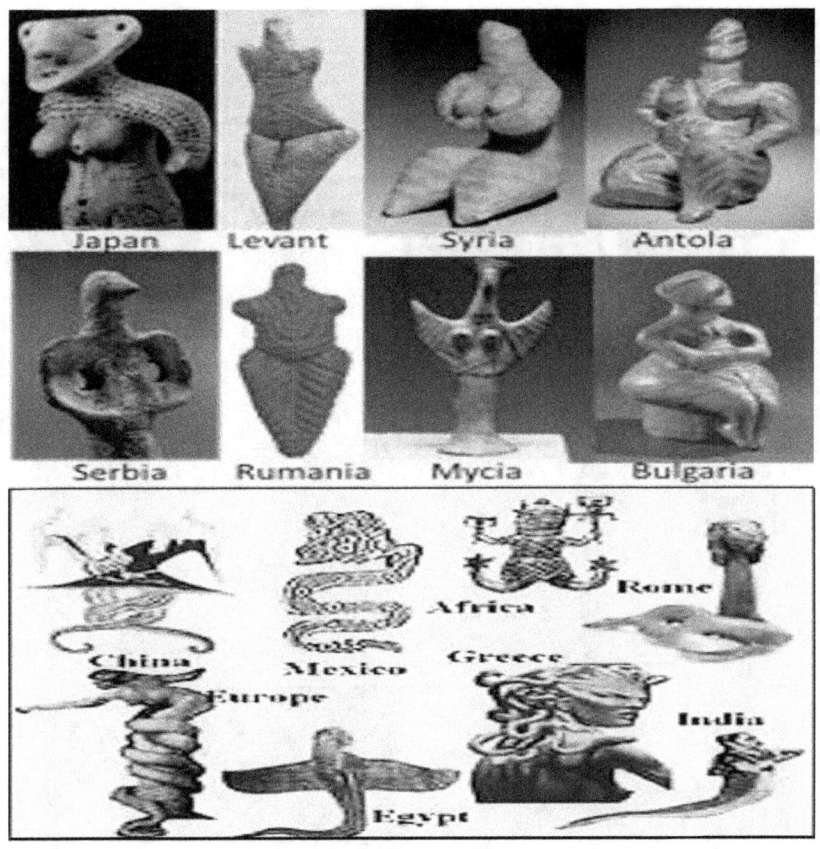

On the left we see lizard people with snakeskin and to the right we see some even had snake tails.

Bumps

I don't know if this is a mark of the mutation or simply a fashion statement of the Lizard mutated people, but many of these people are shown with massive bumps on their shoulders. When bumps are not shown the image of a swastika are painted on. A sampling of these "Bumps" is shown next.

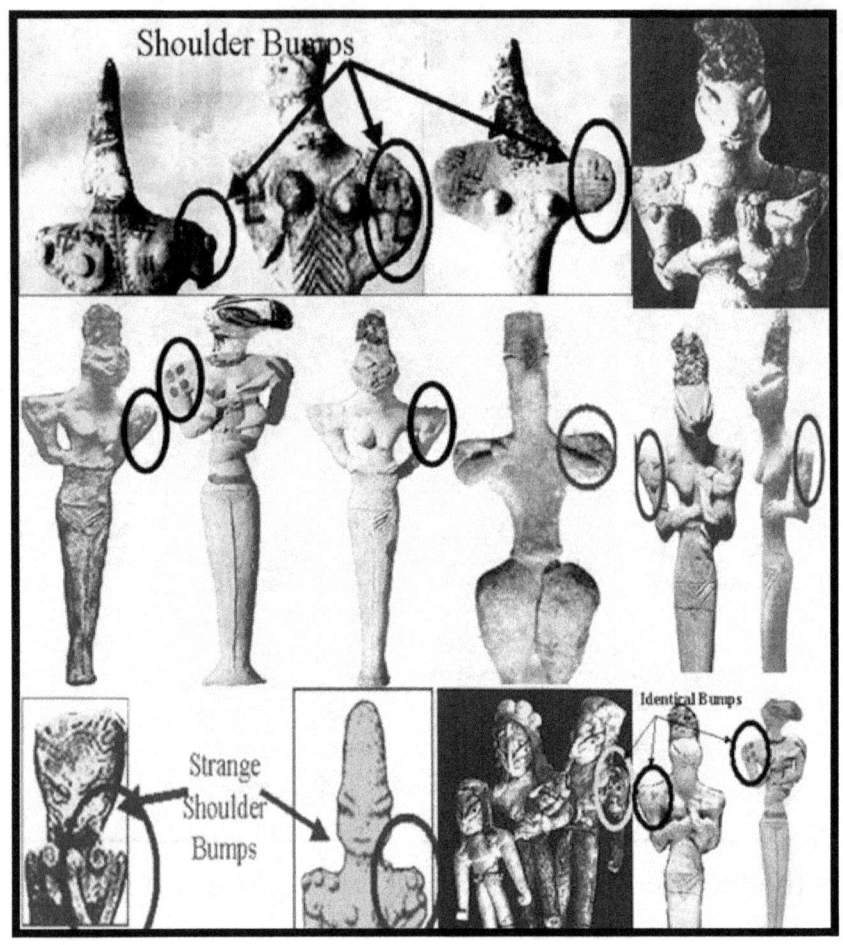

Images and statuettes of lizard people are found all over the world. A small sampling follows.

Persian Lizard-men-All over Persia we find this type of de-volved humans were everywhere.

Indian Lizard-men- The same type people were integrated into societies in India. [Below]

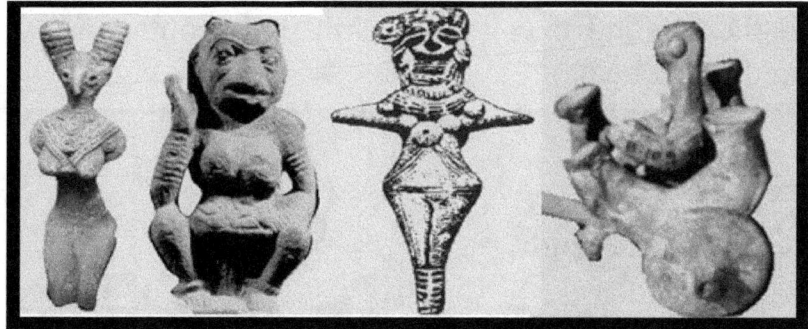

Chineses Lizard men-In China, they were cvalled the Dogu and there are many representations of these unfortunate people. The tiny mouth and large almond eyes show they were a relative race to the others. [Following]

In Eastern Europe, we find the same race of people with the almond eyes, small chin and mouth and what looked like a beak.

In the UK, we find the same race of people again. While we can't tell much about the eyes, the pointed beaklike head is unmistakable.

Aztec Lizard People-Here we find bumps, coffee-bean eyes, and some with broad shoulders just like many of the other similar people from other parts of the world. They were unmistakably different than NORMAL people. The "odd" people lived with "normal" people. The 4 images on the following collage are from this same society.

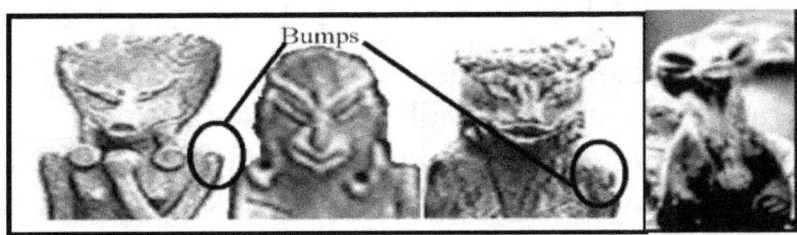

Nigerian Reptilian Gods-This Nigerian Snake God may have legs were like snakes and his face was almost human like. Still this ugly human was considered a god. Even though one would think that a god would never be associated with a snake. This guy was revered by the people of ancient Nigeria.

  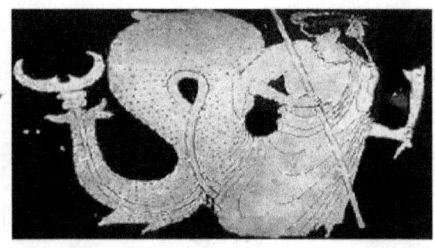

Greek Reptile People-*Gorgon & Nersus*- With her snake hair and awesome power, Medusa and the other Gorgons were possibly depictions of the mutated reptilian people. Nersus was the snake god founder of Greece was most possibly a reptilian person as well.

Indian Reptile Gods-In India, we find the same type of ugly reptilian person. The 4 shown left are some of the female ones.

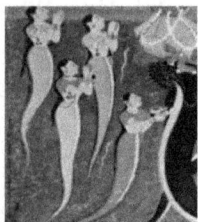  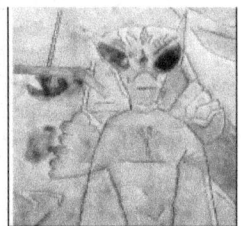

Egyptian Lizard People-The etching shown, preceding middle, was found on a wall from ancient Egypt. There is no mistake that this was no ordinary reptilian and he had an audience with the Egyptian King. The Egyptians also left us with a drawing of their reptilian–like humanoid in as a frontal view showing his large almond eyes, tiny nose, tiny chin, and tiny mouth. He looks similar not only to other reptilian humans, but he also has a similar appearance to what has been reported as the "Visitors" for unidentified flying machines.

Western African Lizard People-I previously discussed the Nomoli that were designed as ape-men, but we find many of them shaped like lizard-men with scaly skin, fangs, lizard heads, and all the rest. A small sampling is shown next. As I mentioned, one legend indicated that *the statues represent the former kings and chiefs of Temne.*

Lizard Person Families-Like I said this group lived with people around the globe, but what I might not have made clear is that the statues made around the world have unmistakable similarities beyond an ugly reptilian face so these are not coincidences. The reptile people that have been seen by many were similar to people except that they had large almond shaped eyes, very broad shoulders, large heads, and are reptilian-like in their facial features. Normally we would not be able to make such claims, but in this case, there is evidence-- a lot of evidence in the form of drawings and statuettes. Notice the bumps again on the right shoulder of the coffee-bean eyed humanoid to the right.

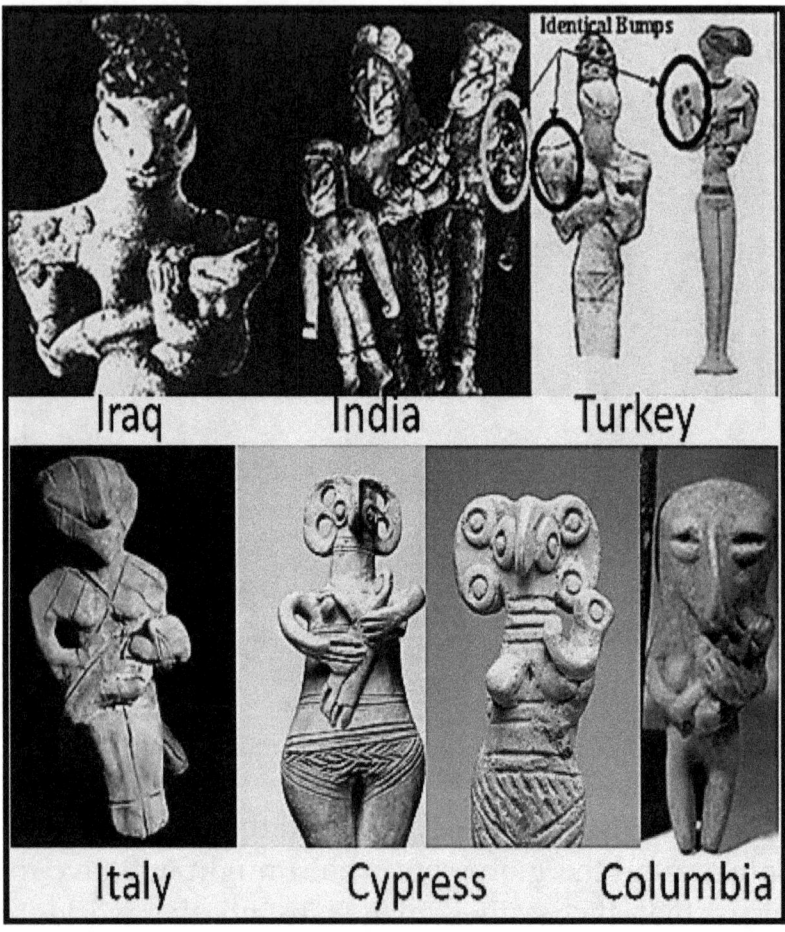

Pakistan & Iraq Reptile Families-The ancestors of the group we will be looking at in that book possibly came from this very same ugly creature. The humanoid depicted, almost assuredly, depicts a real race of people in that identical descriptions and statues of these "people" can be found around the world. To the right are two such examples as depicted about 6 thousand years ago. The family on the middle above, evidently, lived in Mohen-jo-Daro [ancient Pakistan] while the mother and baby depicted on two separate statues on the right came from a very ancient site in Iraq.

For those having a hard time with this mutation as well, all I can say is we still have some people showing up with reptilian features.

Modern Lizard Anomaly

Occasionally we do have some DNA changes that make people more reptilian as is the case with those shown next. Called the Alligator brothers, Swamp Girl, Snake man, and the skin shedding woman, all have reptilian skin that would easily crack if not lubricated, but they did not carry the facial anomalies of the Dogu type humans from long ago.

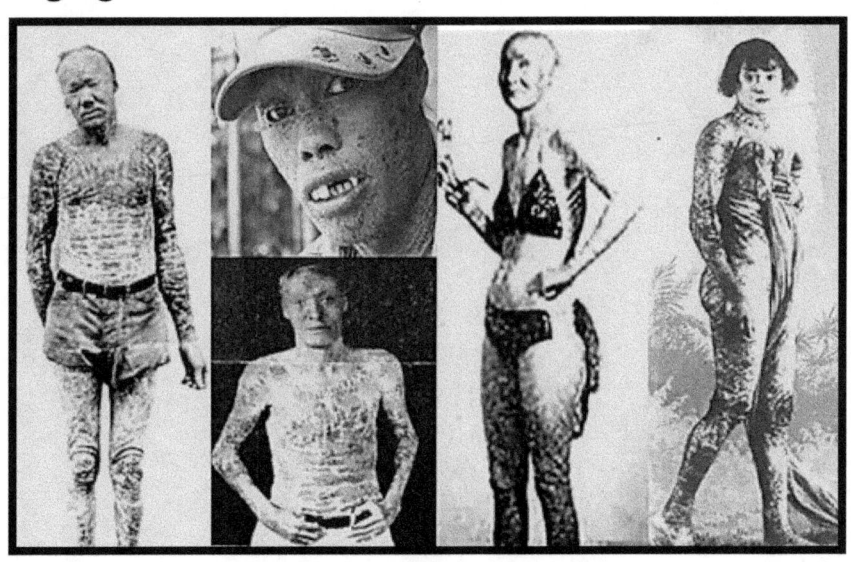

Please know we could easily be changed to a lizard or ape-man with a fine tuning of our DNA or because of radiation modification of DNA. Too much change and we might be a bonobo. While this is a horrible thing to learn, without knowing a truth or even a close proximity to the truth, we will be destined to repeat past errors again and again.

Considerations

I think you have been given a lot of data to consider. Here is an overview of some of the main topics presented concerning how DNA data is being turned intro DNA anomaly so that favored theories are not jeopardized by truth. While there is substantial information concerning the various humanoids, the details have been modified, rejected, eliminated, and ignored. This is no way to help our children. While we only looked at a small sampling of the DNA modification that can help us understand our TRUE history, I hope you got enough of a taste to investigate further and tell you children and get involved with school to push for new histories based on fact rather than desire. The following is a bird's eye view of the main humans and humanoids described in this book and when they lived. If teachers would just talk about these people it would certainly help.

Homo Type	Common Name	Exist [x000 Yrs Ago]	Time on Earth-TYA
Gigantus	Titan	250-120	130
Capensis	Anak	120-3	118
Habilis Types	Australopithicus to Habilis	120-80	40
Erectus Types	Ergaster to Neanderthal	90 to 15	85
Sapien-Sapien	Cro-Magnon	35-5.5	30
Vanara	Ape-men	5.5 to 3.5	2
Ubaid	Lizard-men	5.5 to 3.5	2
Chimp/Bonobo	New Apes	5.5 to now	5.5
Cognatus	Big-Foot	5.5 to now	5.5
Modern	man	5.5 to now	5.5

How can students be expected to understand the really hard elements of living if they are lied to at every step.

The Nuclear Decay timing fiasco has been covered up and still is used to justify uncontrolled evolution so the idea that prehistoric humans modified DNA.

The Prehistoric Nuclear Plant fiasco- To hide the fact that Homo-Gigantus had nuclear energy and most human mutations occurred from nuclear fallout and war this was considered anomaly.

Young Dryas Nuclear Fallout fiasco- They cannot admit nuclear fallout was from war so it is completely anomalized.

Radioactive, Unfossilized Dinosaurs- They cannot admit Pleistocene scientist modified DNA so they are trying to hide this.

Two Races of Humans- While test after test says the same thing, this must be hidden so no one will know most races are false.

Out of Africa Error- This gives uncontrolled Evolution as reason even when it is unreasonable.

Survival of the Fittest- Even knowing the Law of Entropy has not halted this stupidity.

Bonobo from Man Anomaly- This would make it look like a massive nuclear war changed our DNA 5 thousand years ago.

Vanara People Anomaly- This would make it look like a massive nuclear war changed our DNA 5 thousand years ago.

Ubaid Lizard People Anomaly- This would make it look like a massive nuclear war changed our DNA 5 thousand years ago.

Asia America Land-bridge stupidity- This must be pushed so no one would believe in the thousands of depictions of Ancient Aircraft.

X haplotype DNA in America Anomaly- Eliminated so no one would ask about ancient aircraft

T haplotype DNA in Homo- Cognatus and Cherokee- Eliminated so no one would question why 10-thousand-year-old Middle Easterners would be ape-like.

If we can't accept realistic timing, that physical evidence probably has a meaning, that ancient people were much more cleaver than we make them out to be, and that wars have devastated civilizations many more times than World War I and II.

Today we have our 2 "Normal" races of humans, the Homo-Cognatus humans hiding in the woods, and the chimpanzee race. Additionally, some of the people who escaped the Bharata War appear to be coming back from time to time to see if they can reestablish some type of union with the "modern" humans, but that is a different story.

The End

About the Author

Steve Preston is a long lime author of scientific, esoteric facts. His books focus on the painful truths rather than whitewashed details that make us comfortable. If you are interested in the truth instead of comfort, please review other works by Mr. Preston as shown below. The images are some from Egypt taking the older version of taxi similar to what Moses might have used. To the right the writer is shown in the Jewish Negev desert of Israel where the Dead Sea Scrolls were found that were used by John the Baptist in his teachings.

His books include a wide assortment of different subjects including Biblical History and proofs, the story of man's development, Ancient Technology, new views of Physics and Biology, Ancient Wars, current fears and events. A partial list follows.

Development of Mankind

The First Creation of Man-book 1 History of mankind
The Second Creation of Man-book 2 History of mankind
The Creation of Adam and Eve-book 3 History of mankind
The Antediluvian War Years-book 4 History of mankind

Man After The Flood-book 5 History of mankind
Close Look at Ancient History-book 6 History of mankind
A New View of Modern History-book 7 History of mankind
The Twentieth Century and Beyond- Book 8 History of Mankind

Bible History, Correction, and Analysis

Abraham to Moses-First part of the Bible
Adam's First Wife-Story of Lilith
Adam to Abraham- Second Part of the Bible
Closer Look At Genesis- 200 ancient text confirm Genesis
Exploring Exodus- Reviewing the Details of "Exodus"
Errors in Understanding- Interpretations of the Bible
Expanded Genesis- Apocrypha and other Jewish texts
Exploring Genesis- Reviewing the details of "Genesis'
Incarnations of God- How often did God become Incarnated?
History Confirmed By The Bible- Science confirmation of the Bible
Moses Saved Egypt- How the Jews eliminated the Hyksos
Moses to Jesus- Third part of the Bible Series
Mysteries of the Exodus- Proofs of the Exodus
New look at the Bible- Questions in Interpretation
Old Testament Used By Jesus- Ancient Jewish texts
Understanding the New Testament-4th part of the Bible Series
Why the King James Bible Failed- Issues with KJB

Ancient Technology and Life

Anakim Gods- History of the Ancient Giant/gods
Ancient History of Flying- Ancient flying
Kingdoms Before the Flood- Pleistocene humans
Living on Venus- Venus before the Pleistocene Extinction
Martians- Ancient Life on Mars
Mysterious Pyramids- Who made the Pyramids?
Victory of the Earth- History of our Earth
Not from Space- UFOs are not from space.

Amazing Technology- Descriptions of prehistoric capabilities

Ancient and Modern War

America's Civil War Lie- Truth about the Civil War years
Behind the Tower of Babel- Story of the Bharata War
Driven Underground- Fear in the Bharata War
Four Armageddons- The 4 major wars that destroyed mankind
Six Deaths of Man- Destructions of mankind
World War Before- The Pleistocene War
World War with Heaven- The Angel and Anak War
World War Zero-The Bharata War
When Giants Ruled the Earth- History of the Titan Giants
Sex Crazed Angels- What caused the Heaven War?

Current Events and Fears

Allah' God of the Moon- Terror of Muslims
American School Disaster- fear in our country
Can We Save America? - Fear in the USA
Scythians Conquer Ireland- A History of Ireland
Fast History of MILES Training- Laser based Army training
Great American Quiz- Unusual details of American History
Make Your Own Global Warming
Truth About Phoenicia- The Evidence -First in America
Monsters are Alive- Post Pleistocene Monsters
Promote the General Welfare- Fear in USA
Our Very Odd Presidents- President review
Terror of Global Warming- Fake issue uncovered
The Antichrist- Many demonic possessed rulers
The Bad Side of Lincoln- Negative side of a great man
The Devil- Of Demons and their master
Vampires among Us- How Demons and Vampires are similar
Humans on Display- Slavery and Human Zoos

New Look at Physics

Amazing Technology- Pleistocene Technology
Anthropic Reality- We control our Reality
Consensus Science- Fake Science
Complex Earth- Truth behind Earth's development
Is Time Travel Possible? Science of Time Travel
Retiming the Earth- Eliminate of Nuclear Decay Errors
Releasing Your Consciousness- Beyond our SELF
Slip Through a Wall- How to walk through solids
Our 12-Dimensional Universe- New science of our Universe
Mystery of Photons and Light- Science of Photons
Of Heaven and Hell- scientific descriptions
Meaning of Life and Light- Detains of New Science
Vibrational Matter- New Science of Quantum Fluctuations

New Look at Biology

DNA of Our Ancestors- Tracing DNA of ancient man
God Didn't Make the Ape- New science on ape Evolution
Lizard People- Mutated People of the Bharata War
Creation and Death of Dinosaurs- Why Dinosaurs died
Races of Men- Tracing DNA of Humans
Tracing Cro-Magnon to Jesus- The third creation and mutation
Self, Soul, Spirit- Three components of Life
Self-Virtualization- New science of reality
True Happiness- Self Actualism and Beyond
Life Resonance- Unusual capabilities of men
Awaken the Departed- We can talk to the Dead
Biophotonics and Healing- How Photonics used in medicine

A Look at Anomaly

Religious Anomalies- Looks at Answers to Biblical questions
Flying Anomalies- Looks for answers to flying questions
DNA Anomalies- Looks for answers to genetic questions
Planet Anomalies- Looks for answers to genetic questions